Version 5.0

Automated Homework System

Quant Systems, Inc.

Charleston, SC

ISBN 0-918091-04-7

Table of Contents

An Overview of the *Adventures* Concept

Adventures in Statistics is a tutoring system designed to enhance the understanding of statistical concepts, improve problem solving skills, and make the process of learning more enjoyable. The *Adventures* system is designed to provide tutoring upon request and allow the student to test themselves in a "fear free" environment. *Adventures* relies on the mastery level learning concept. That is, the student may take a test (certification) as many times as is necessary to demonstrate mastery of the material. There is no penalty for failing during the learning process--after all, failure is an important part of learning. We would like to think that the only way to "fail" any module is to quit trying.

Typically, instructors assign *Adventures in Statistics* modules as they would assign homework from a book. Each module tutors and tests students on one statistical concept. Since problem scenarios, parameters, and key wordings are randomly generated, each student's problem set is unique.

Adventures problem solving exercises are designed to be used in your school's computer lab or at home or work on an IBM PC compatible. Each *Adventures* module begins in the learning mode where the student works with our intelligent tutor. After developing confidence in your ability to solve problems, switch to the certification mode (testing) and try to demonstrate mastery of the topic. Once the module is successfully completed, a certification code is automatically logged and time-stamped into our database, the Classroom Management System. Students who are working at home can bring the certification code to the school's network and register it into the database. If a student does not pass the test, he/she can retake the test until mastery of the topic is demonstrated and the code is received.

Installation Instructions for *Adventures in Statistics*

If you have any questions or if we can be of assistance in any way, please don't hesitate to call us at (803) 571-2825.

Minimum System Requirements

1. MS DOS 5.0 or higher
2. Windows 3.1 or higher
3. 20 MB of hard disk space
4. 4 MB RAM (8 MB highly recommended)

Installation Procedure

a. Insert the diskette labeled "*Adventures in Statistics* Setup Disk #1" into the floppy disk drive and then run MS Windows (if you are not already in Windows).

b. Select **File** (Alt + F) from the Program Manager main menu and then select **Run** (Alt + R) from the pull-down menu.

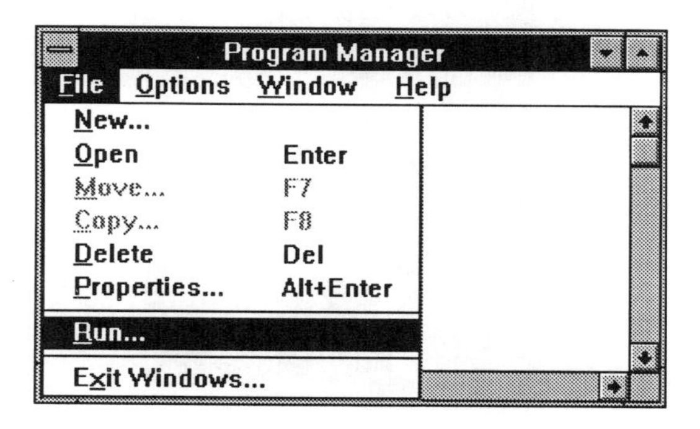

c. Type the path to the floppy drive in which you inserted the installation diskette and then type **setup**. For example, if you inserted the setup diskette into drive **a:**, type the following:

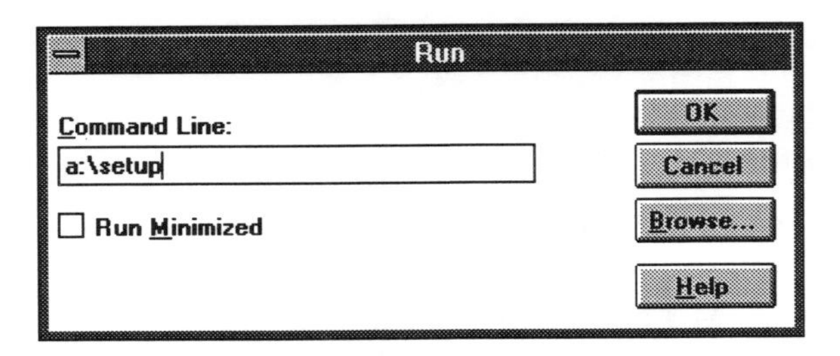

d. Press <Enter> or click on the **OK** button.

Specific to Windows 95:

a. Insert the diskette labeled "*Adventures in Statistics* Setup Disk #1" in the floppy drive.

b. Select Start and then select **Run** (Alt + R).

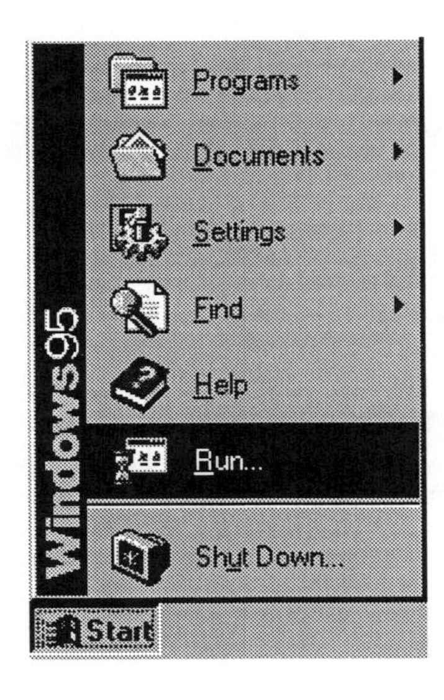

c. Type the path to the floppy drive in which you inserted the installation diskette and then type **setup.exe**. For example, if you inserted the setup diskette into drive **a:**, type the following:

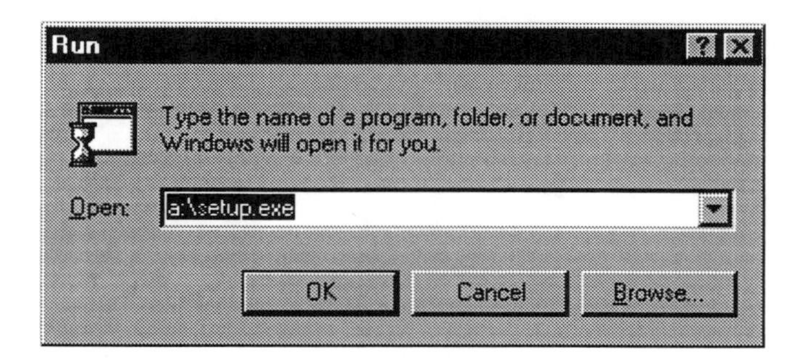

During the setup process you will be asked to do the following:

a. Type your name and your institution's name. Then select **Continue Setup**.

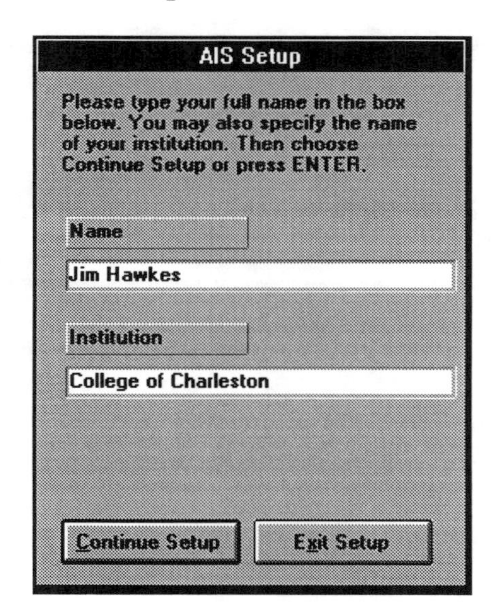

b. Confirm your input and then select **Continue Setup**.

c. Enter the path to the directory and the name of the directory in which you would like to install *Adventures in Statistics*, and then select **Continue Setup**. In the following example, *Adventures in Statistics* will be stored in the AIS directory on the hard drive **c:**. (Note: The default is c:\AIS.)

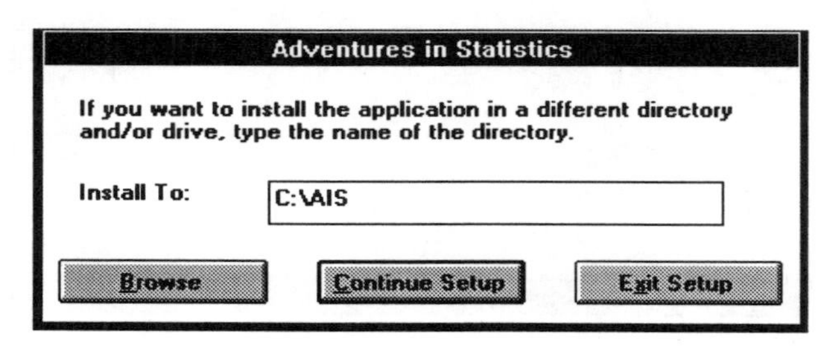

d. After selecting **Continue Setup** in part (c) , a box will appear displaying the progress of the setup. If you would like to quit the setup program at any time, select **Exit Setup**.

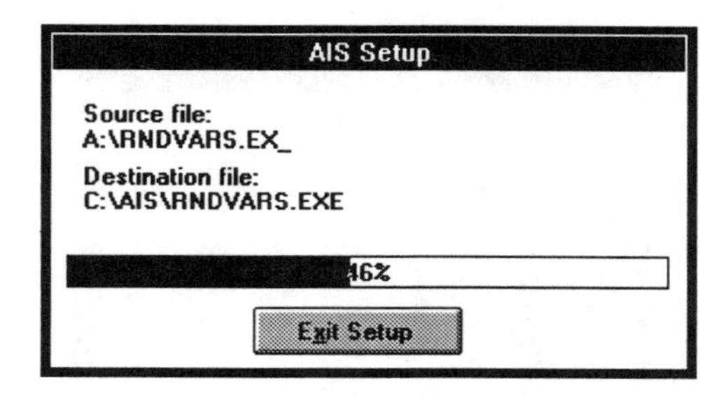

e. During the setup process, you will be prompted to insert more disks. Insert the disks, in the order that they are requested, into the same drive that you inserted Setup Disk #1. Select **OK** after each disk is inserted.

f. Type the name that you would like to call the *Adventures* program group in the box below and then click on the **OK** button. (Note: The default is *Adventures in Statistics*.)

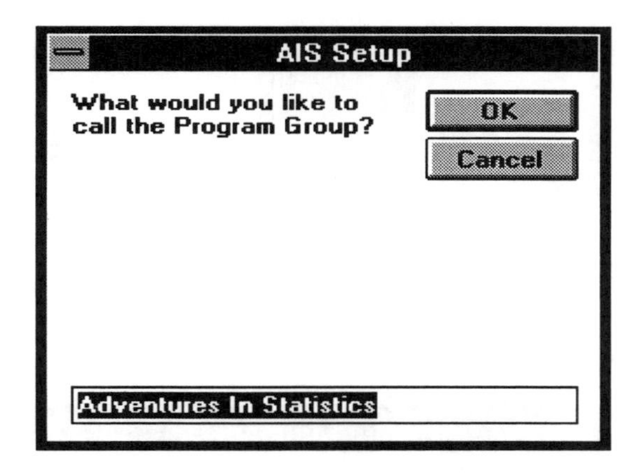

g. The setup program will create a program group called *Adventures in Statistics* (or the name you entered).

Note 1: Windows creates the group file name ais<u>name</u>.grp for the group, where name=the first 5 characters of the name given to the program group.

Note 2: The installation procedure is the same whether you are installing the program onto an individual computer or onto a network. However, for network installations, if the individual workstations do not access Windows from the network drive, the program group will have to be set up on each individual workstation.

Starting *Adventures in Statistics*

1. Run Microsoft Windows.

2. Open the *Adventures in Statistics* program group by double-clicking on the *Adventures in Statistics* icon.

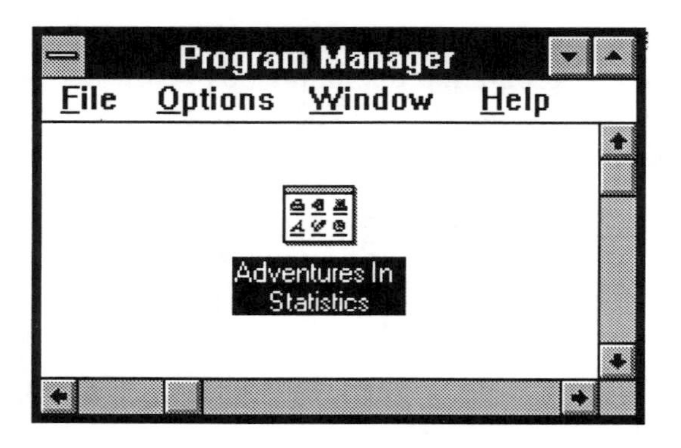

3. The *Adventures in Statistics* program group will appear. To start *Adventures in Statistics,* double-click on the *AIS Index* icon.

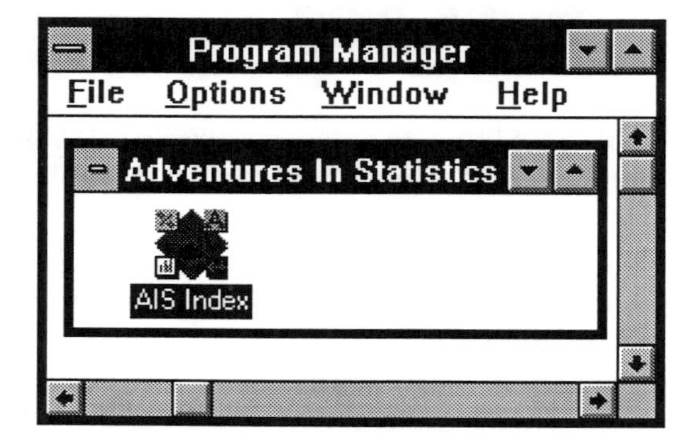

Specific to Windows 95:

Select **Programs** from the **Start** menu. In the **Programs** menu select *Adventures in Statistics* and then select *AIS Index.*

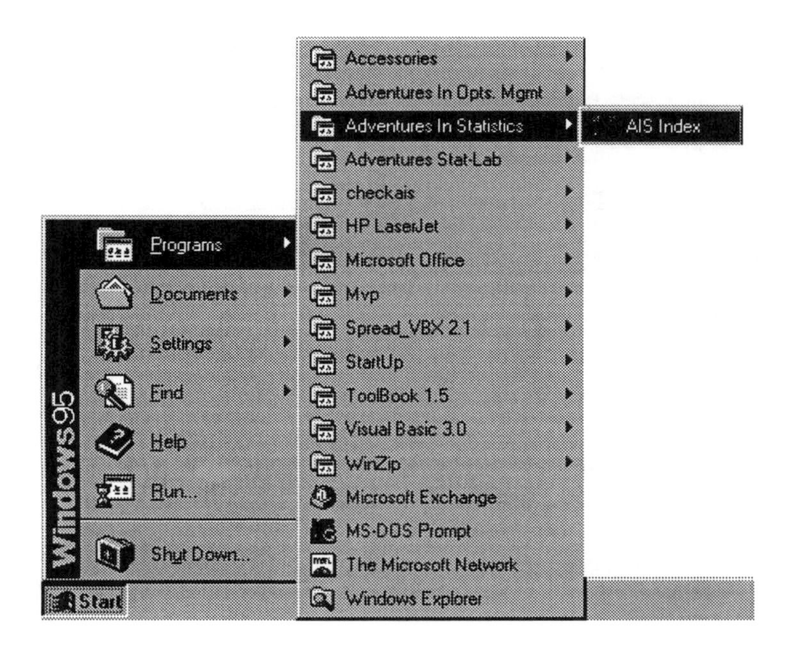

4. The *Adventures in Statistics* menu will appear.

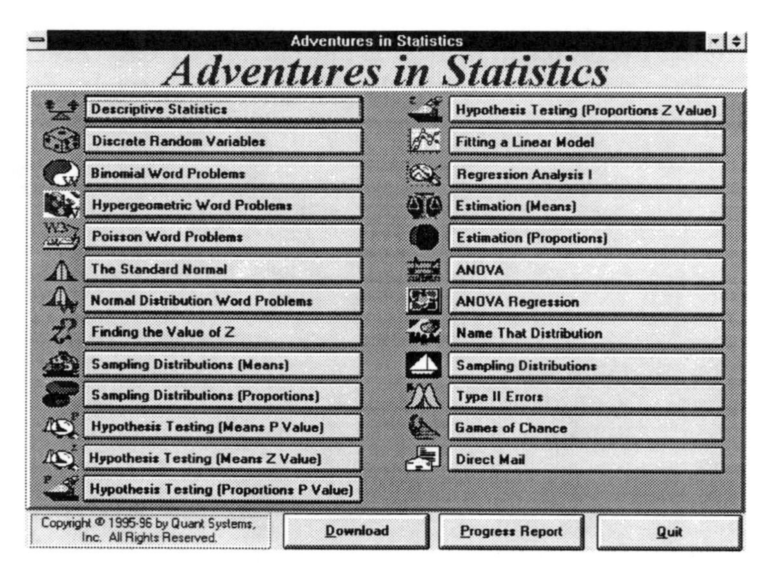

a. To download *Adventures in Statistics* from the network so that you can work at home, select **Download** (Alt + D).

b. To see a report of your progress or to register your certification code, select **Progress Report** (Alt + P).

c. If you wish to exit *Adventures in Statistics*, select **Quit** (Alt + Q).

5. Selecting a Module

From the *Adventures in Statistics* menu, click once on a module name button. (Or press the up and down arrow keys until the desired module is highlighted and then press Enter.) After a module is selected, the title screen for that module will appear.

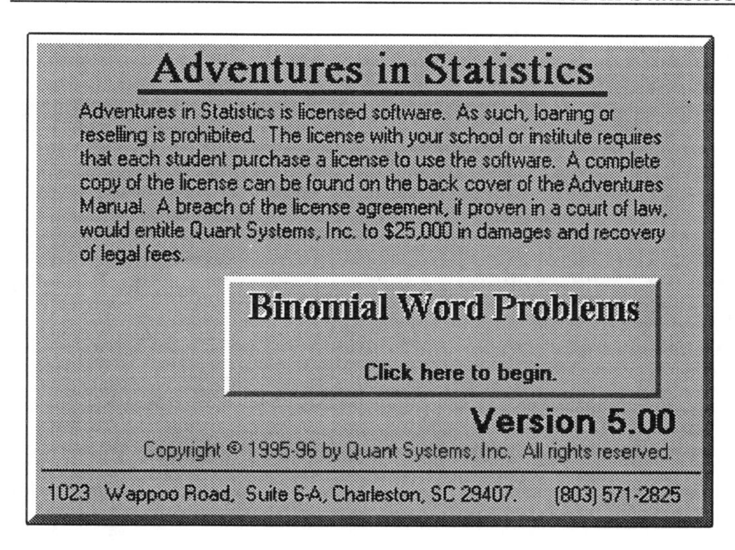

Click one time on the module name button. The authorization code entry box will appear.

6. **The Authorization Code**

Students will receive personal authorization codes from their instructor once they have completed and turned in the coupons that are attached to their manuals. The authorization code is a 30 to 36 digit alphanumeric code that uniquely identifies each student.

The program will request an authorization code each time a module is opened. Students may either type in the code by hand, load it from a disk, or load it from a network drive. If the code has not already been saved on a disk or in a file on a computer, then the code must be typed in at least one time.

7. **Typing in Your Code**

Type each 6 digit section of your code in the spaces provided in the code entry box. After typing in a section of the code, move to the next section by pressing <Enter>. After typing in the last section of the code, either press <Enter> or click on the **Ok** button.

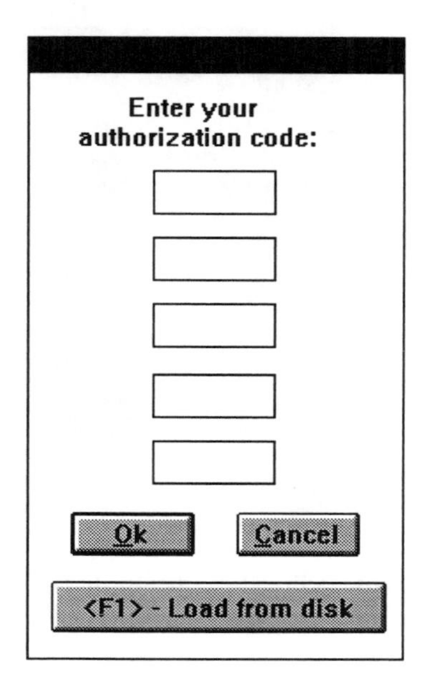

Cautions:

1) Make sure to type the **letter I** (as in igloo) when your code contains this letter. Authorization codes do not contain the **letter L** or the **number 1**.

2) Make sure to press <**Enter**> after each section of code. Your code may not work if you press <**Tab**> instead.

8. **Saving Your Code**

After you type in your code and it is accepted, the program will ask you if you want to save your code to a disk. You can save your code on a floppy disk or on your computer's hard disk.

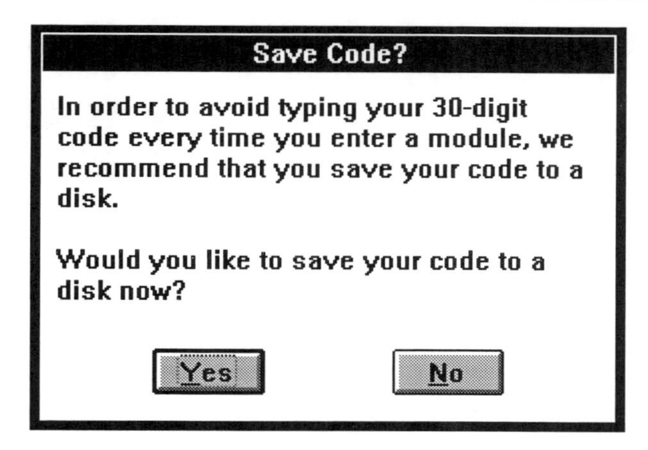

If you do not want to save your code now, select **No**. You will have to type in your code every time you open a module until you save your code.

If you select **Yes**, the program will ask you to choose the drive and directory on which you wish to save your code.

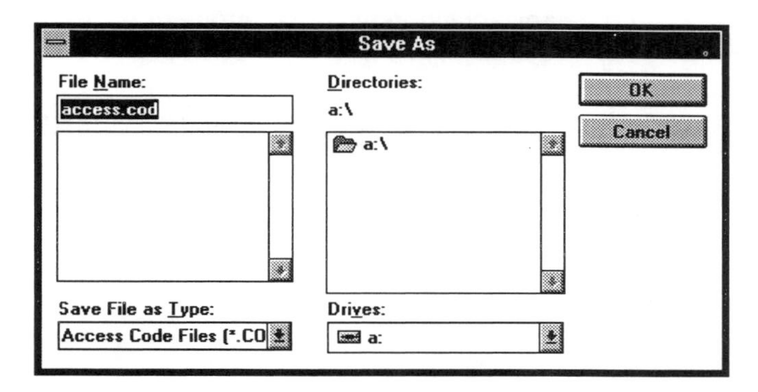

In this example, the code is being saved to a file named **access.cod** that is in the **root directory** of a floppy disk in **drive a:**. After you choose the drive and directory, click on the **OK** button or press <Enter>.

9. **Loading Your Previously Saved Code**

When the code entry box appears, press the <F1> key or click on the **Load From Disk** button.

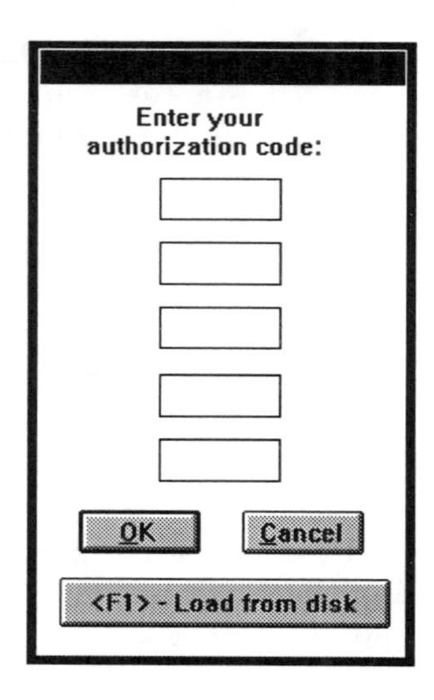

From the Open box, choose the drive and directory where your code is located. Then either double click on the access.cod file name, click on the **OK** button , or press <Enter>.

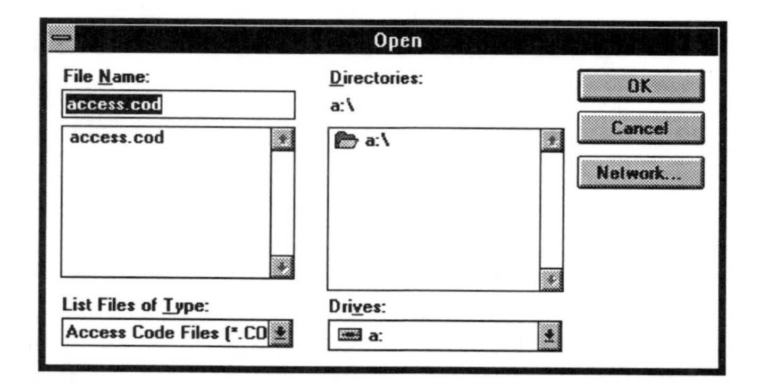

When your code has been accepted, the following box will appear:

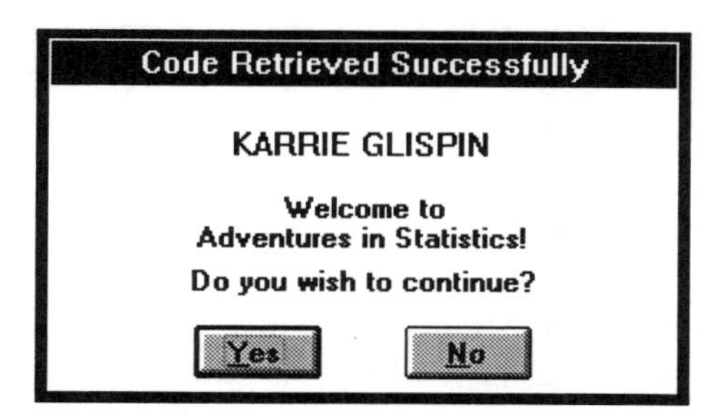

Select **Yes** to continue.

10. Next, you will be asked to select the Series # that was assigned by your instructor.

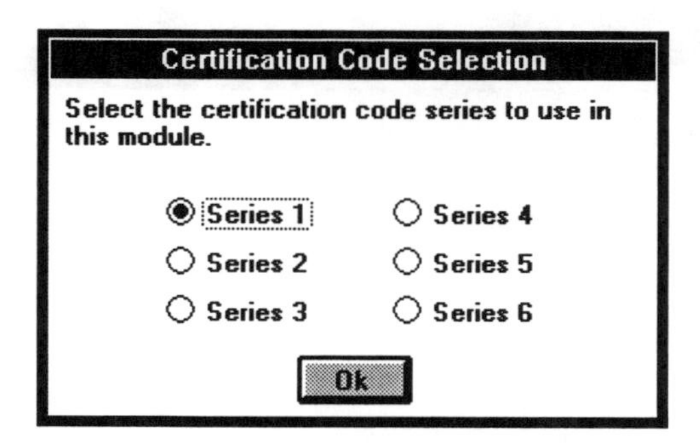

Choose a Series # and then select **Ok**. You can now proceed with the module.

Note: Unless your instructor specifically tells you otherwise, leave the series set at "Series 1". If your instructor is using the Classroom Management System (CMS) and you are working on the school's network, the series is set by your instructor and you will not be prompted to select a series code.

Modes of Operation

Most of the *Adventures in Statistics* modules have a Demonstration Mode. Every module has a Practice Mode and a Certification Mode.

Demonstration Mode

The Demonstration Mode highlights the module's unique features by walking the student through a sample problem. The step by step instruction was created so that instructors will not have to spend class time showing students how to navigate each module.

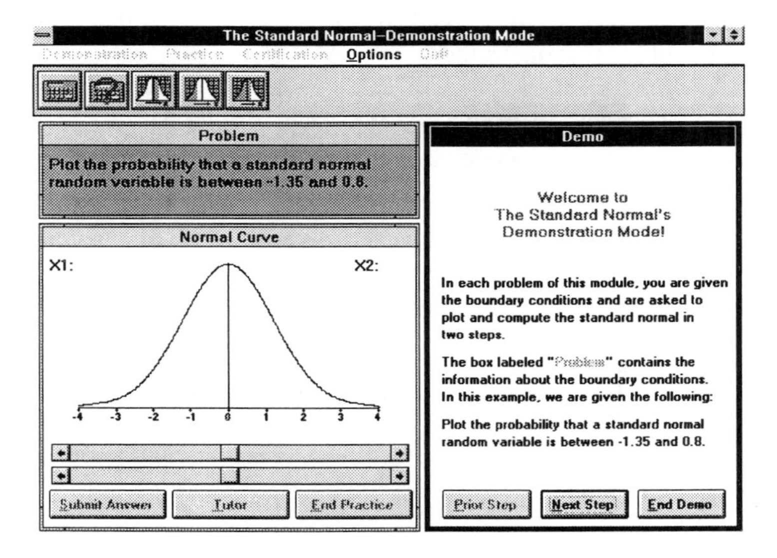

The Demonstration Mode features the following information:

1. the type of problems that students will be expected to solve,

2. the mechanics of using the module's unique features, and

3. the criteria for mastery.

Practice Mode

In the Practice Mode, help is provided through the "Tutor" and "Help" options. The "Help" option is used to describe the mechanics of inputting an answer. The "Tutor" option gives a discussion of the topic, detailed examples, a detailed solution to each randomly generated problem, and hints on solving the problem. If a student makes an error, some modules will provide an explanation of the error.

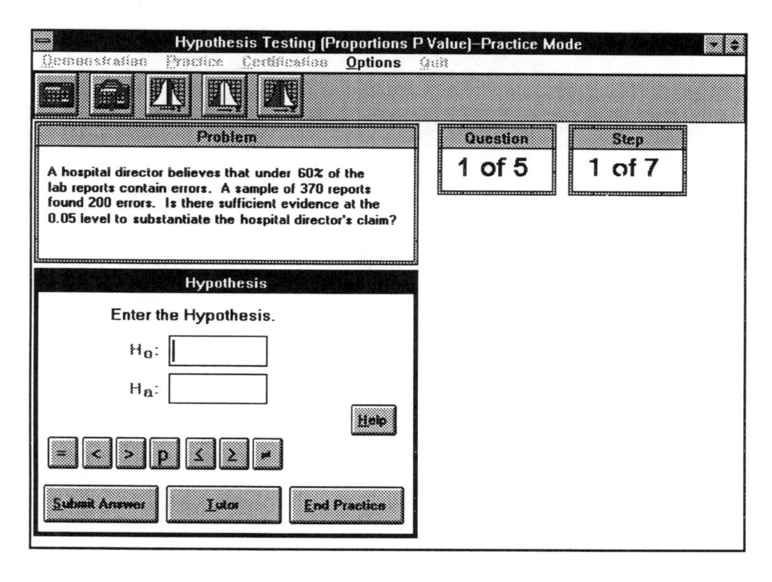

Certification Mode

The Certification Mode is a testing mode in which the tutoring function is removed. In this mode, students must correctly answer 80% to 90% of the questions in order to "certify". The percentage required to certify varies by module. If students are unsuccessful in their first attempt to certify, they may try again as many times as is necessary to master the concept.

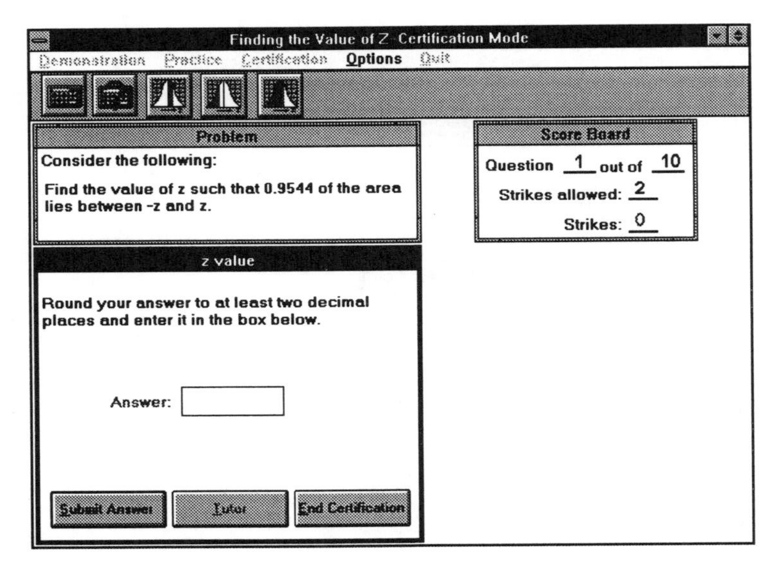

Once students successfully complete a module, the program produces a five or six digit certification code that is unique to each student and to each module. Students may want to record this code in the Record Book that is in the back of this manual.

The certification code is either submitted physically to the instructor or electronically through the Classroom Management System.

Commands and Conventions

In documenting the *Adventures* software, we use angle brackets <> to enclose a key that is to be depressed. Do not type the angle brackets.

Selecting an Option or a Button

If you wish to select an option (or a button), click the mouse on the option (or the button), or hold down the <Alt> key and press the key of the underlined letter at the same time.

Example: To select the **Practice** option on the horizontal menu bar, click the mouse on the word **Practice,** or hold down the <Alt> key and press <P> at the same time.

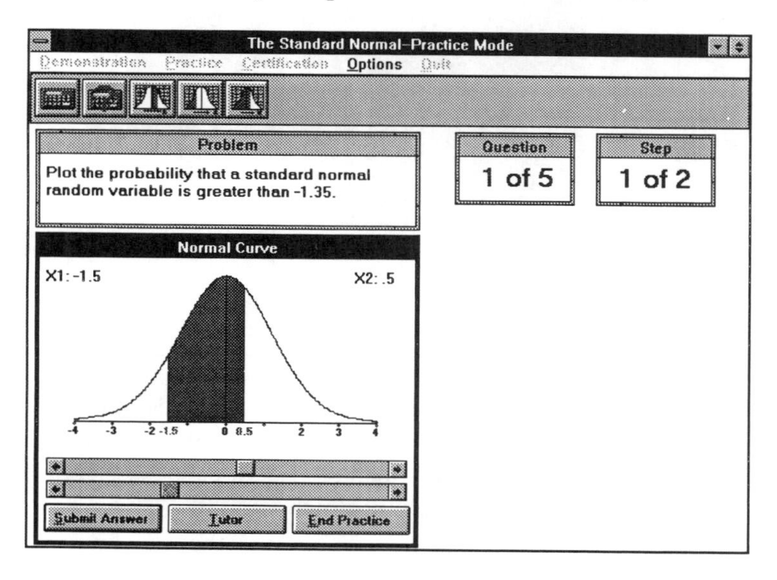

Example: If you wish to select the **End Practice** button, click the mouse on the button, or hold down the <Alt> key and press <E> at the same time.

The Submit Answer Button

When in the Practice Mode and the Certification Mode of most *Adventures in Statistics* modules, select the Submit Answer button to enter a complete answer.

The Tutor Button

The **Tutor** button is only available in the Practice Mode. The **Tutor** gives access to the **Lecture**, **Hint**, and **Solution** options.

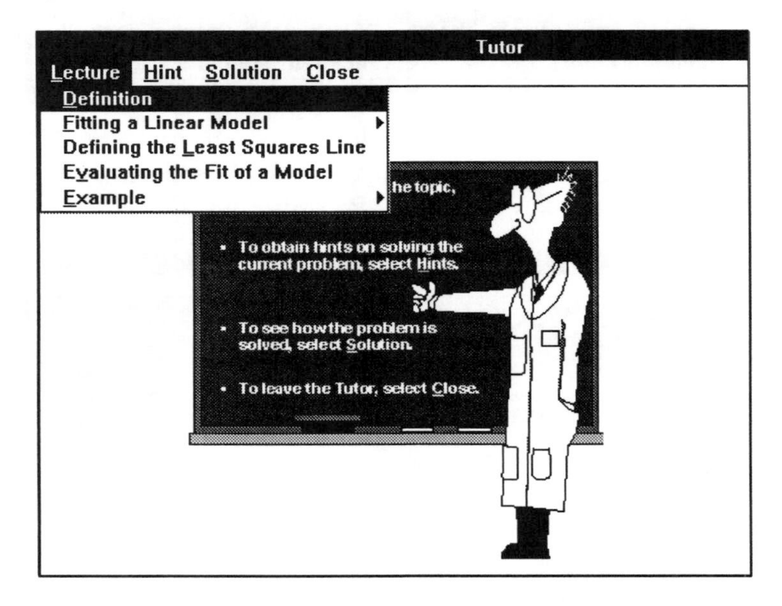

The Lecture is designed to help students understand the concepts of the module. It also gives example problems with detailed solutions. To open the Lecture, select **Lecture** and choose a page from the pull-down window.

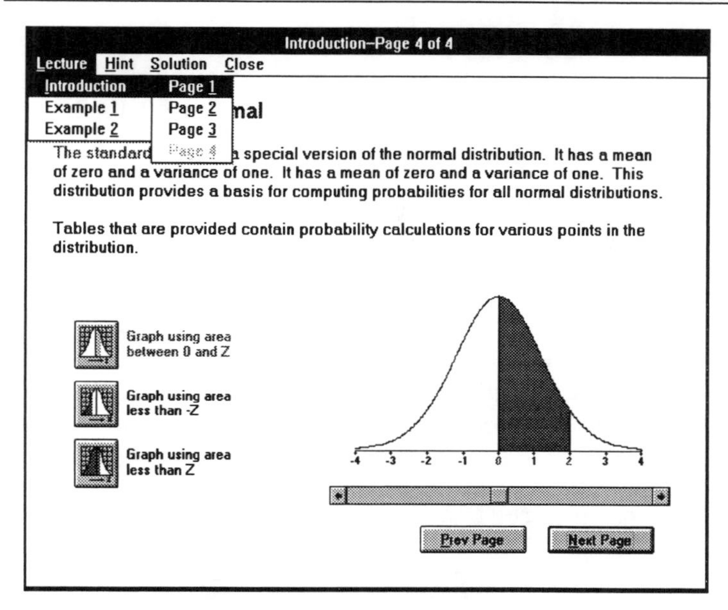

Move through the Lecture by selecting the **Next Page/Previous Page** buttons, or go directly to a page by selecting **Lecture** and then choose the desired page from the pull-down menu.

If the **Solution** option is selected from the Tutor screen or from one of the Lecture screens, students will be shown in detail how to solve the given problem, and then they must move on to the next problem.

If the **Hints** option is selected from the Tutor screen or from one of the Lecture screens, students will be shown small steps that will gradually lead them toward the solution of the given problem.

What Happens When an Incorrect Answer is Submitted?

In all modules where an incorrect answer is submitted, the tutor gives the student two options.

In the Practice Mode, one of the options will always be **Try Again**. A student can try to answer the given question as many times as desired. This option is not available in the Certification Mode.

In the Certification Mode, one of the options will always be **Display Answer**. If this option is chosen, the correct answer will be displayed in red.

The second option given when an incorrect answer is submitted will either be **Explain Error** in the Practice Mode and in the Certification Mode, or it will be will be **Display Hint** in the Practice Mode and **Display Solution** in the Certification Mode.

The Explain Error Button

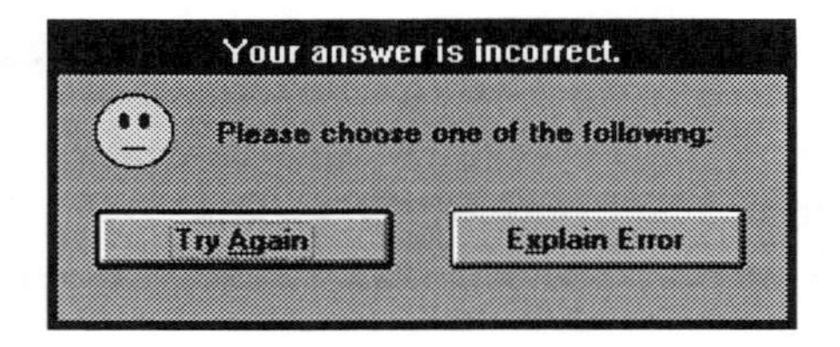

If **Explain Error** is selected, the program will give an explanation of the error that was made.

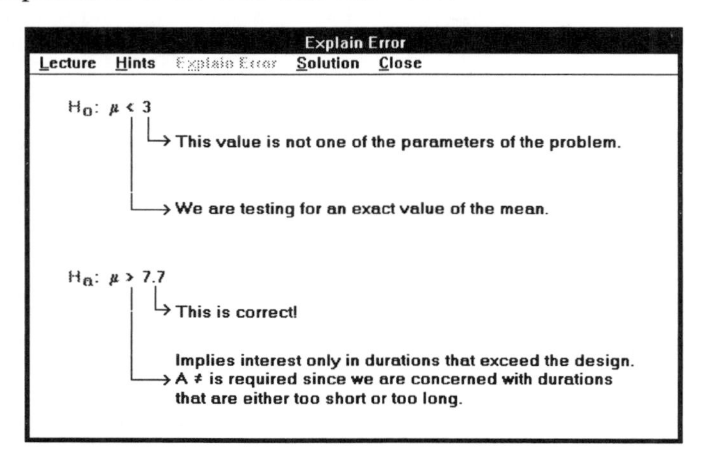

The Display Hint Button

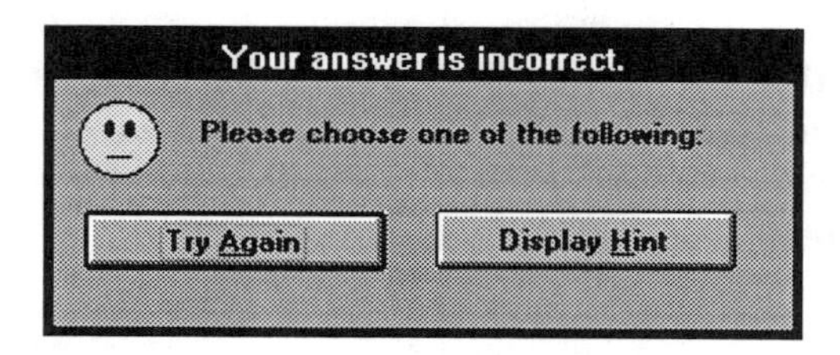

If **Display Hint** is selected, the program will display successive hints to help the student solve the problem.

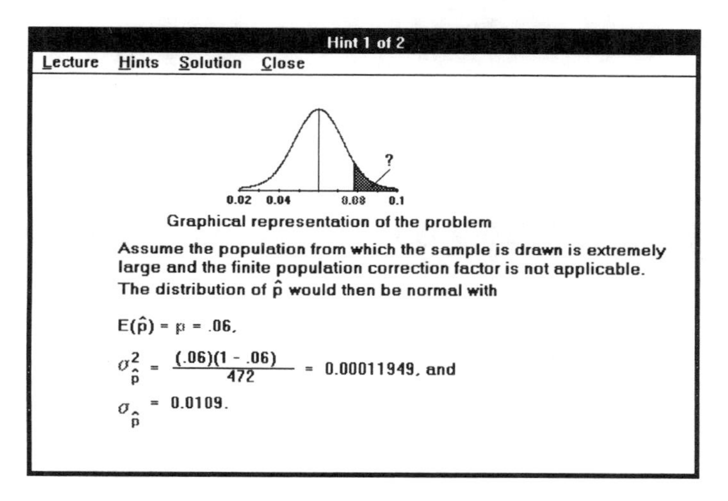

After the last hint is displayed, **Display Hint** changes to **Display Solution**.

Menu Options on the Explain Error and Hint Screens:

The **Lecture** and **Hints** options are only available in the Practice Mode. If **Lecture** or **Hints** and then **Close** are selected, the student can try to answer the question again.

The **Solution** and **Close** options are available in the Practice Mode and in the Certification Mode. If **Close** is selected in the Practice Mode, the student can try to answer the question again. If **Close** is selected in Certification Mode, the correct answer is displayed in red. When **Solution** is selected in either mode, the student is shown in detail how to solve the problem that was given. Then the student must move on to the next problem.

The Display Solution Button

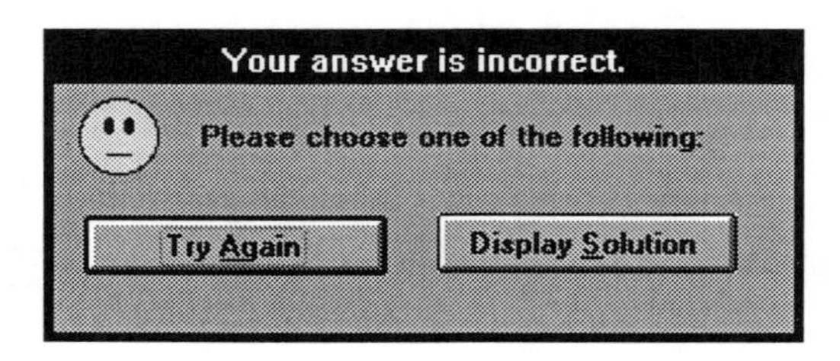

If **Display Solution** is selected, the program will show in detail how to solve the problem that was given.

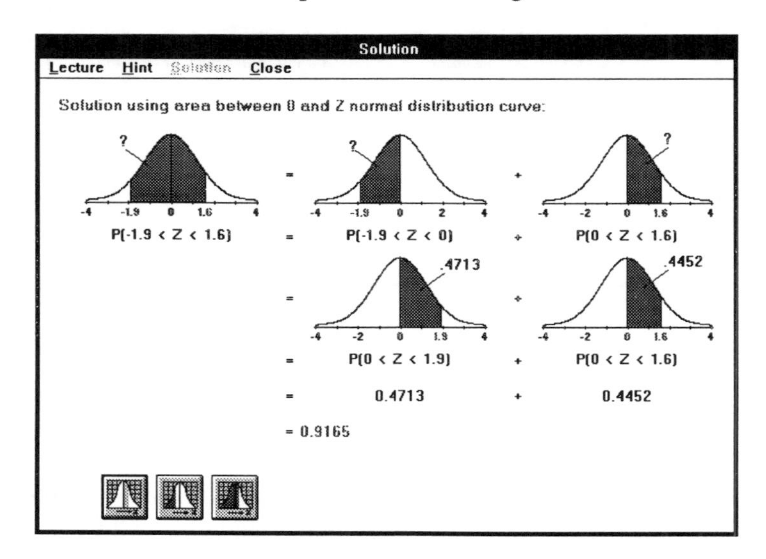

Menu Options on the Solution Screen:

The **Lecture** and **Hints** options are only available in the Practice Mode. If **Lecture** or **Hints** and then **Close** are selected, the correct answer is displayed in red.

The **Close** option is available in the Practice Mode and in the Certification Mode. When **Close** is selected, the correct answer is displayed in red.

The Help Button

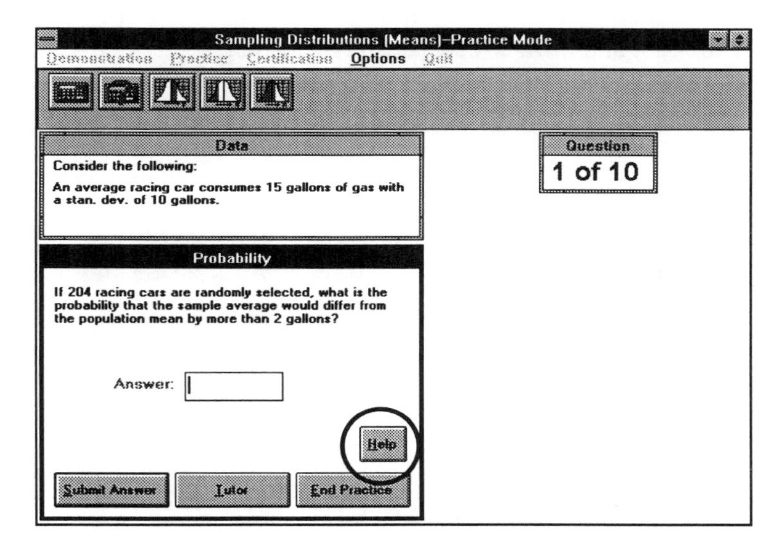

When necessary, a **Help** button is available that gives detailed explanations of how an answer should be entered. It also gives directions on how to use the tool buttons available in the module.

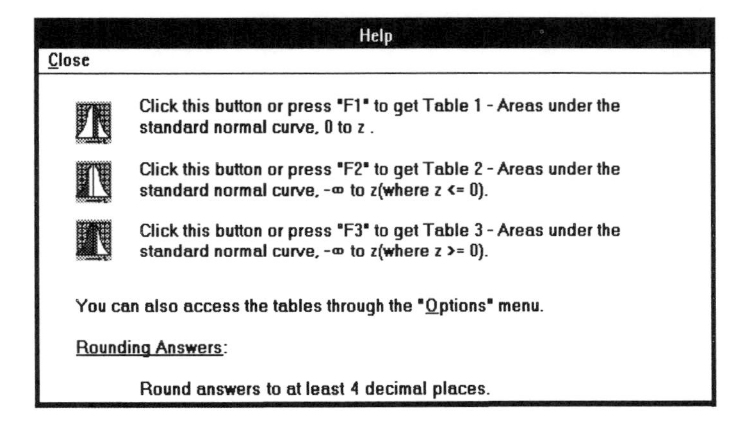

The Toolbar

Several modules have a toolbar available under the horizontal menu bar. Two tools that will be on most toolbars are the Calculator and the Calculator Help. If a tool has a related help tool, it will be located directly to the right of the tool and have a red question mark on it.

Other tools that can be found on the toolbar will depend on the module. For example, the toolbar in the Sampling Distributions (Means) module has three tools that display the three normal distribution tables.

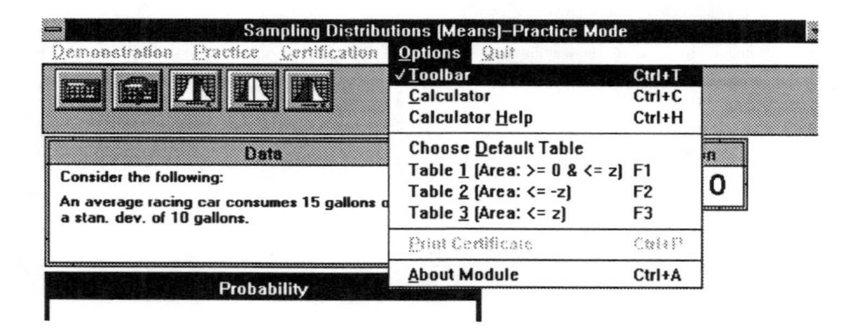

All of these tools can be accessed by clicking on the tool buttons with the mouse. They can also be accessed from the **Options** pull-down menu.

Classroom Management System (CMS)

The Classroom Management System is a database that allows students to enter their certification codes into the instructor's grade book. The system also allows students to view their personal progress reports and the instructor's syllabus.

Enrolling in CMS

If CMS has been installed, students entering an *Adventures* module or the Student Report Program for the first time will be prompted to enroll in CMS.

First, they will be asked to select their professor and enroll in the appropriate section.

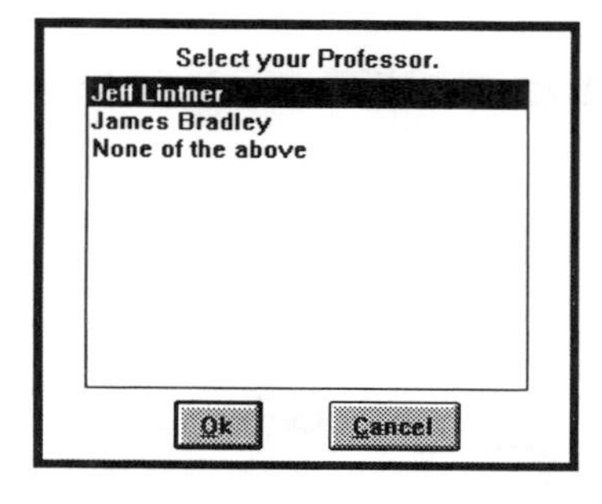

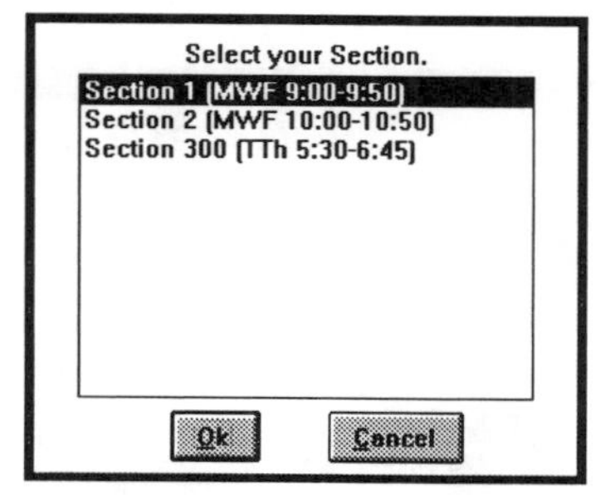

Once the selections have been made, the students will be prompted to confirm their selections.

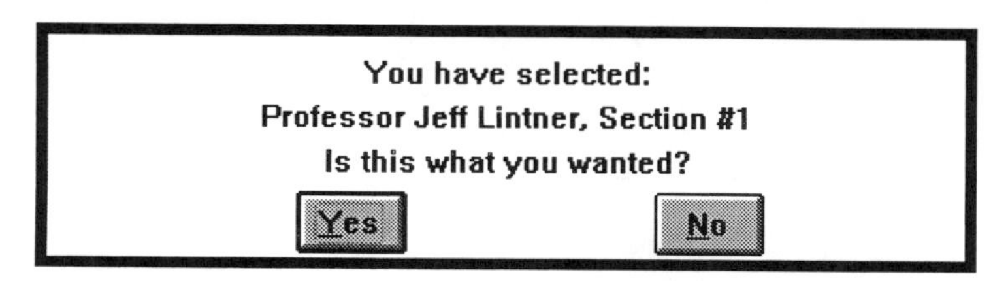

The student's selections will be acknowledged with the following box.

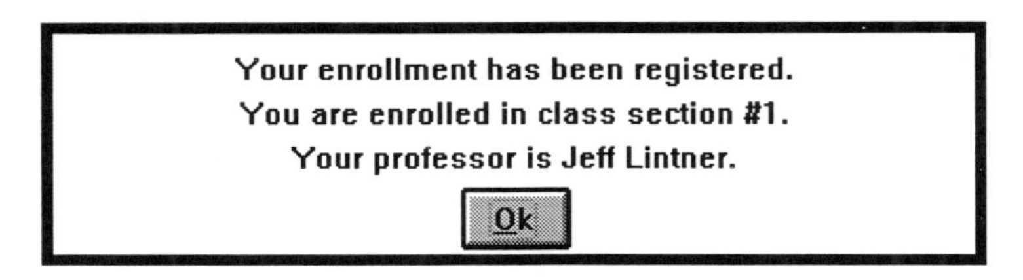

After the students have enrolled in an instructor's section, they will not be prompted to enroll again.

The Student Report Program

During the installation of CMS, a button called "Progress Report" will be added to the *Adventures in Statistics* Index menu. Students will use this program to enter their certification codes and review their progress.

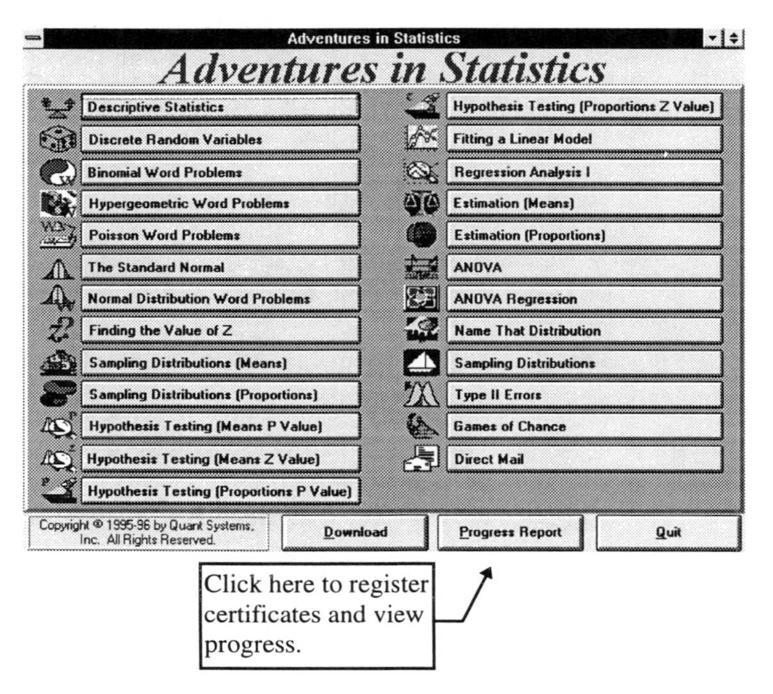

Registering a Certificate

Once students have certified in a module, they can register their certificates in the instructor's grade book by performing the following steps.

a. Select the **Progress Report** button from the AIS Index Menu.
b. Enter the access code.
c. Select: **Register Certificates**.

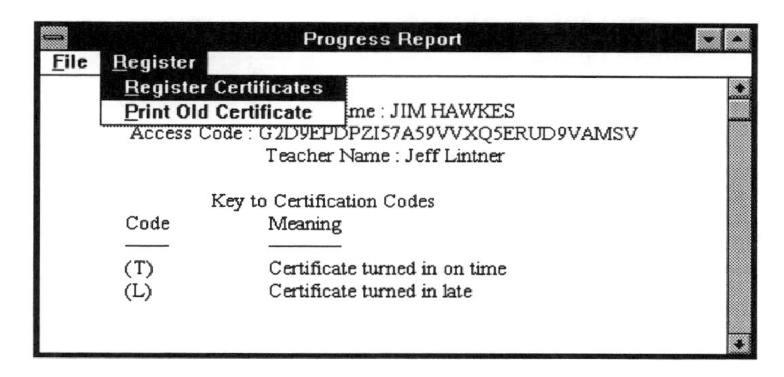

d. Select the appropriate module and then select: **Register**.

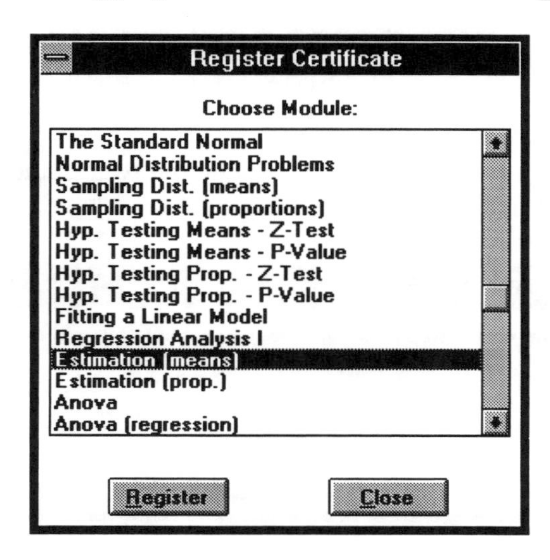

e. Enter the certification code for the module and then select: **<u>O</u>K**.

e. If students enter the correct certification code for the module, they will be notified that the certification has been registered.

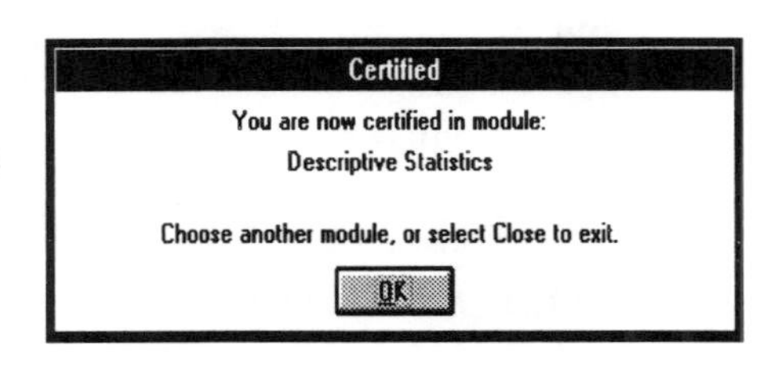

Note: If an invalid code is entered, students will be notified that the code is invalid.

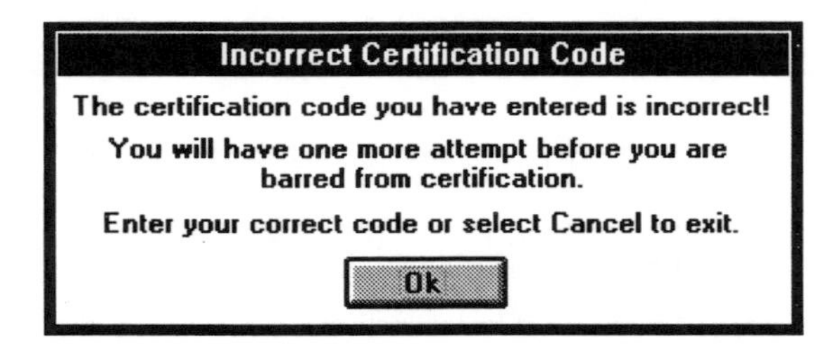

If an invalid certification code is entered twice, students will be barred from certifying in the module.

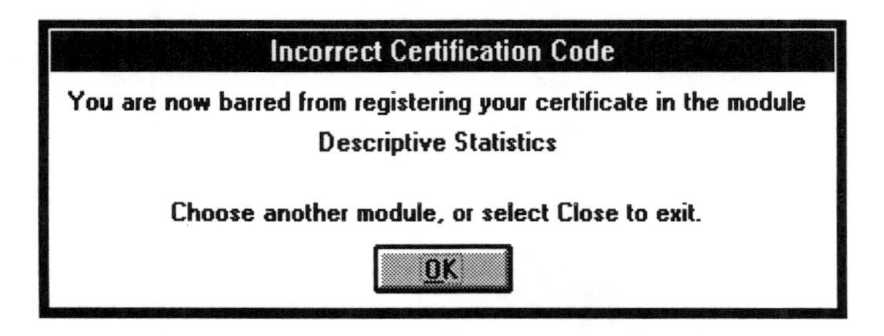

Viewing a Progress Report or the Instructor's Syllabus

Viewing a Progress Report

Students may view a report of their progress to date by selecting the **Progress Report** button from the AIS Index Menu. The progress report will be displayed automatically.

If the report does not all fit on one page, the students can click on the arrows of the scroll bar to move the report up or down.

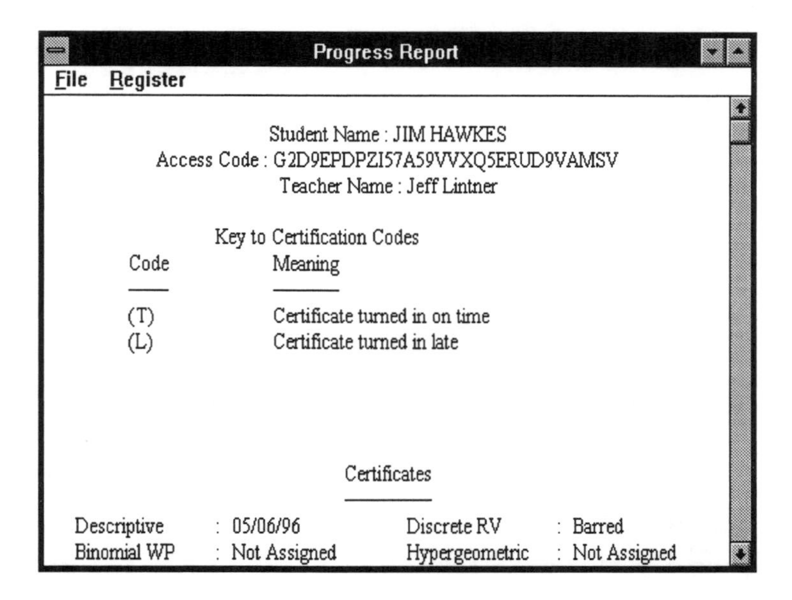

Viewing the Instructor's Syllabus

If the instructor has placed his or her syllabus in the CMS database, students may view the instructor's syllabus by selecting: **File**, **Display**, **Syllabus**.

If the instructor has not placed his or her syllabus in the CMS database, the Syllabus selection will be disabled (gray).

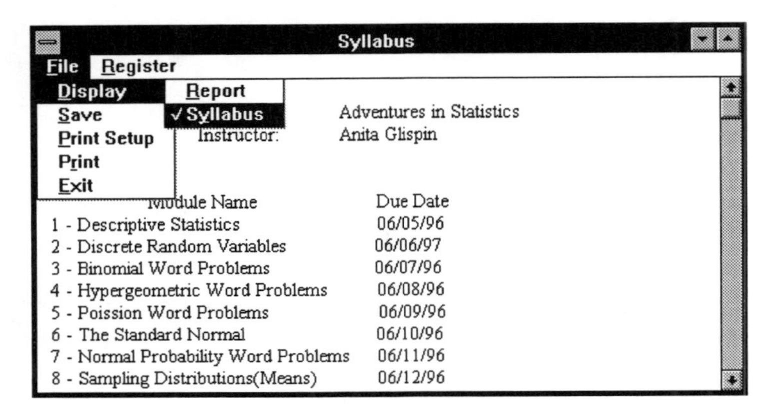

If students would like to view their progress report after viewing the syllabus, they must select: **File, Display, Report**.

Printing or Saving a Progress Report or the Syllabus

While viewing a progress report or the instructor's syllabus, students can print the report or the syllabus by selecting: **File, Print,** or they can save it by selecting: **File, Save**.

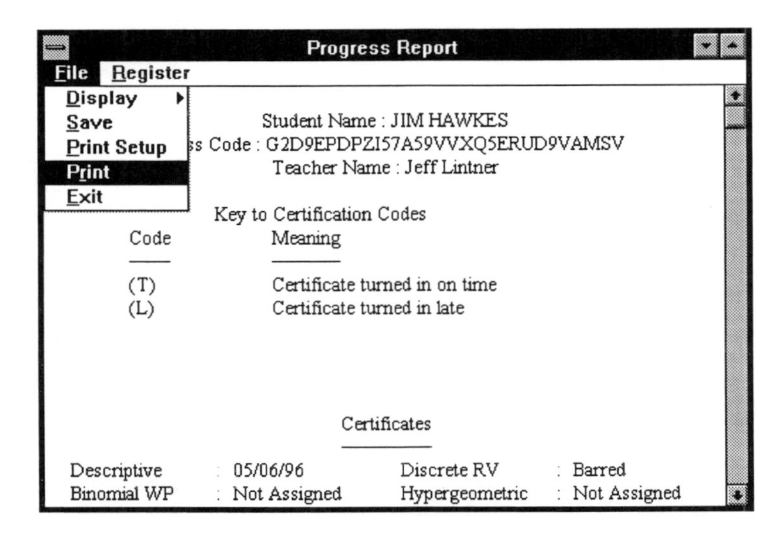

Trouble Shooting

Problem: When you select a module from the *Adventures in Statistics* menu and enter into one of the modes, the following message comes up:

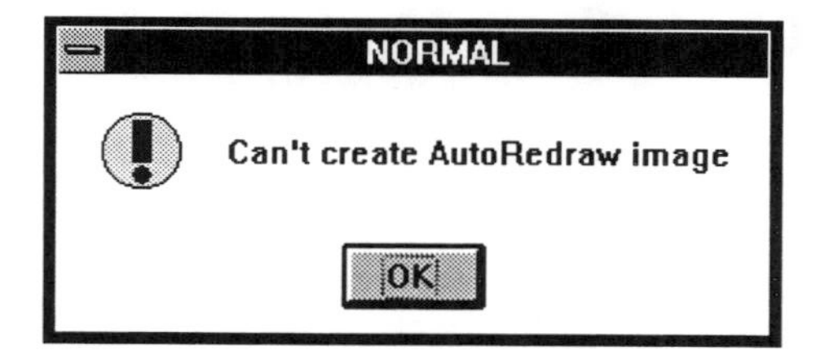

Solution: Increase the size of the Virtual Memory using the following procedure:

a. Open the *Main* program group by double-clicking on the *Main* icon in the program manager.

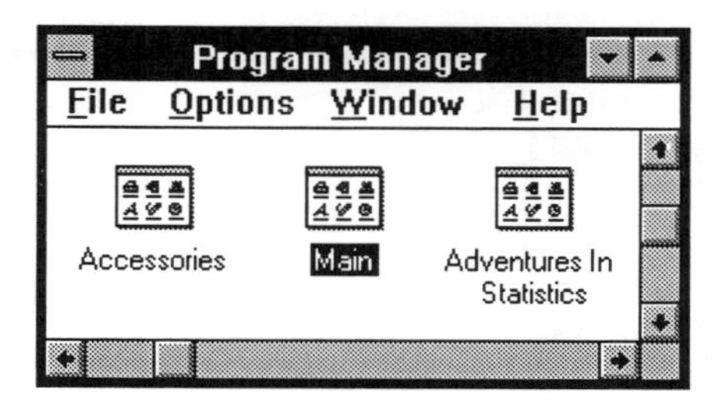

b. The *Main* program group will appear. Open the *Control Panel* program group by double-clicking on the *Control Panel* icon.

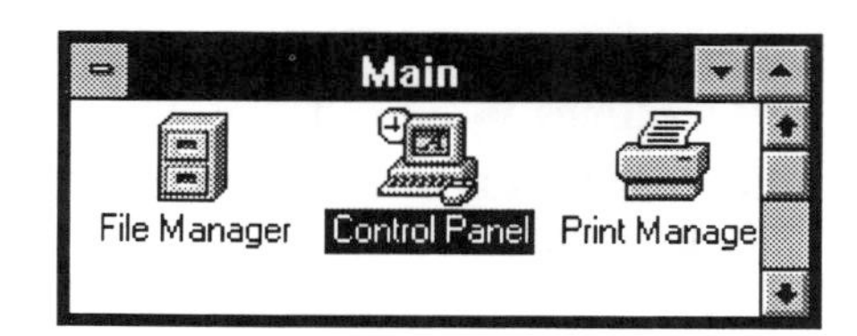

c. In the *Control Panel* program group open the *386 Enhanced* dialog box by double-clicking on the *386 Enhanced* icon.

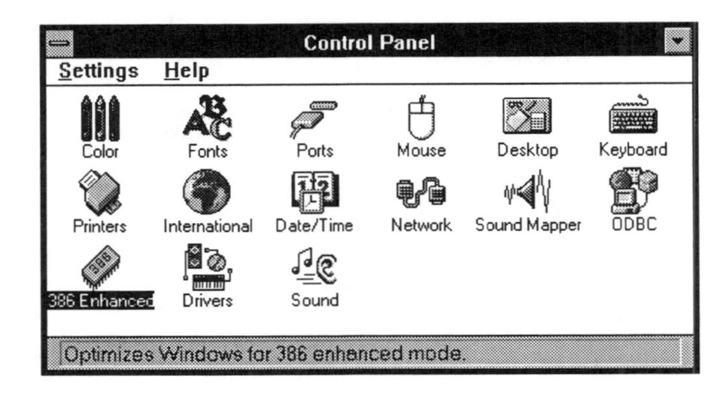

d. In the *386 Enhanced* dialog box, click on the **Virtual Memory** button to open the *Virtual Memory* dialog box.

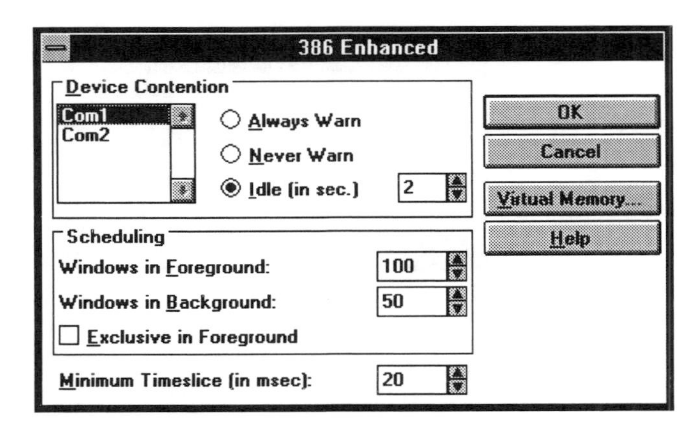

e. In the *Virtual Memory* dialog box you can see the present settings. To change the *Virtual Memory* settings, click on the **Change>>** button. In the following example the *Virtual Memory* size is 2,047KB.

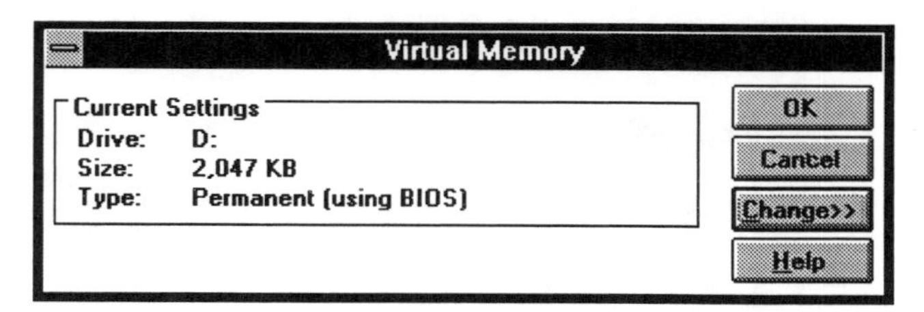

f. Enter the new size in the new size box provided. In the following example the *Virtual Memory* size has been increased to 19,696KB from 2,047KB.

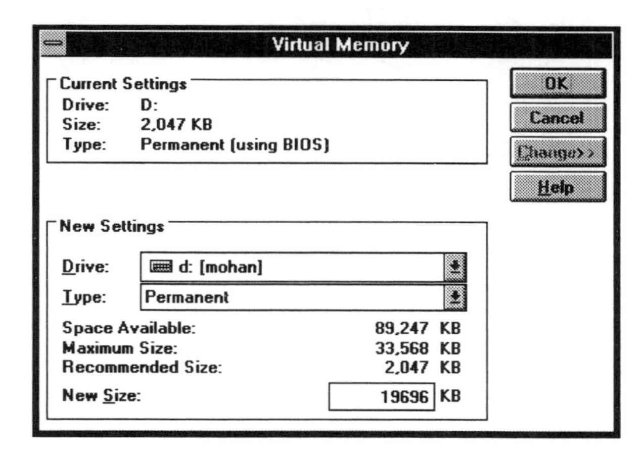

g. Press <Enter> or click on the **OK** button.

h. Confirm that you would like to make changes to *Virtual Memory* settings by selecting **Yes** (Alt + Y).

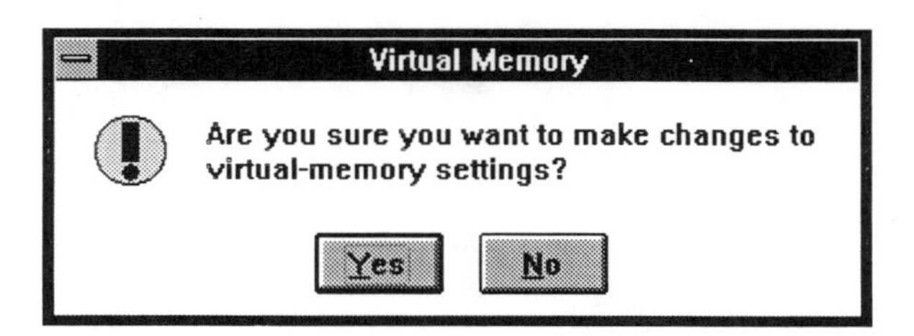

i. The following message will appear:

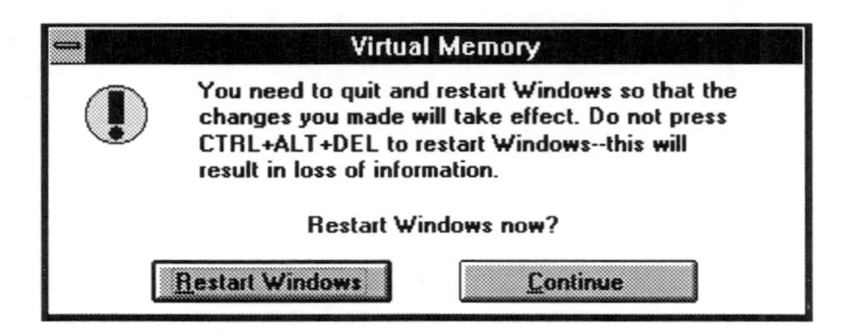

j. Click on the **Restart Windows** button to restart Windows so that the changes made to *Virtual Memory* settings will take effect.

Problem: While you are using a program, a General Protection Fault (GPF) occurs due to a conflict with the display device drivers. The message displayed will be similar to the following:

BINOMIAL caused a general protection fault in module CIRRUS.DRV at 0001:B148

Solution: Change the display device driver to VGA using the following procedure:

a. Open the *Windows Setup* program group by double-clicking on the W*indows Setup* icon in the *Main* program group.

b. Choose the **Change System Settings** in the **Options** menu of the *Windows Setup* dialog box.

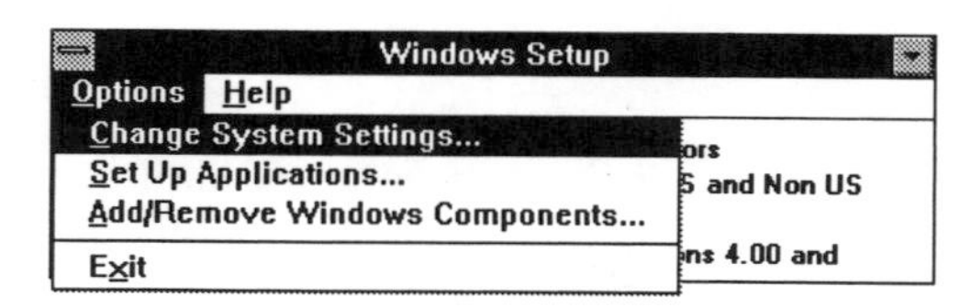

c. The *Change System Settings* dialog box will appear. Choose the option **VGA** in the **Display** list box as shown below.

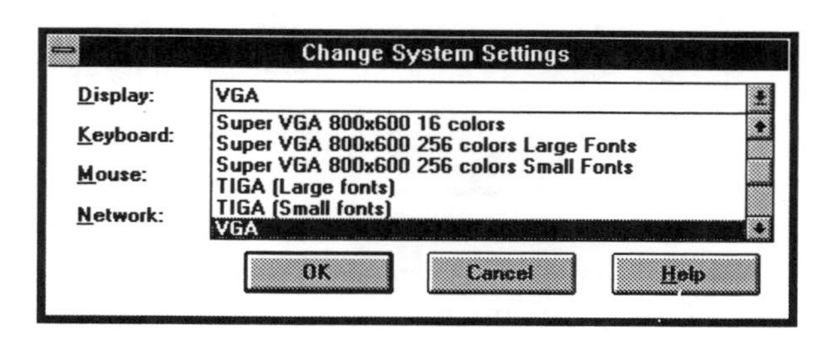

d. The following message box will appear. Click on the **Current** button to use the currently installed driver.

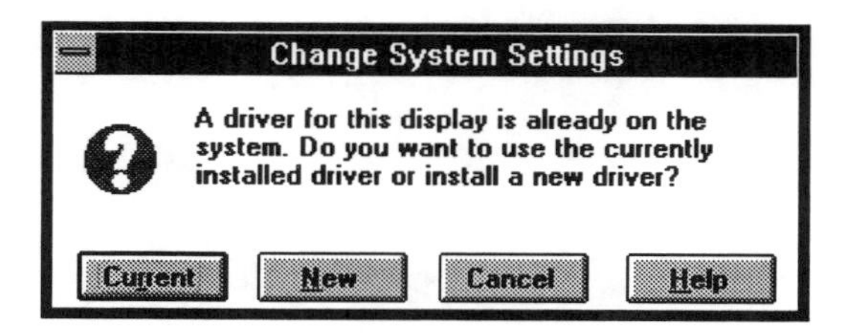

e. Click on the **Restart Windows** button to restart windows so that the changes made will take effect.

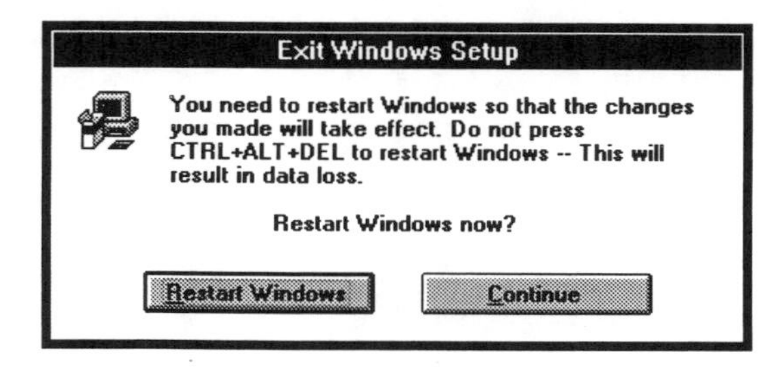

Specific to Windows 95:

a. In the *Control Panel* program group, open the *Display Properties* dialog box by double-clicking on the *Display* icon.

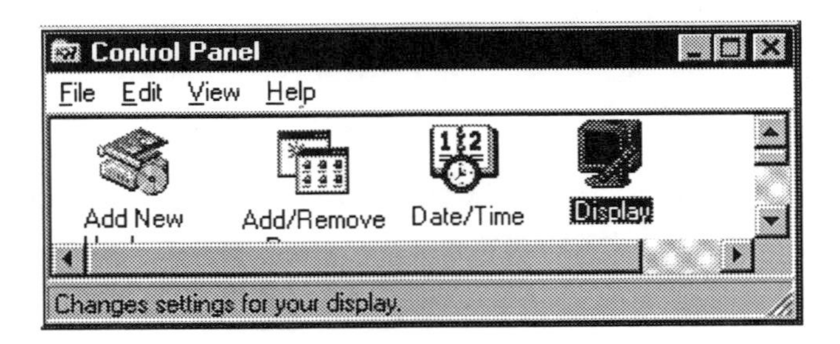

b. In the *Display Properties* dialog box, select **Settings** and then click on the **Change Display Type** button.

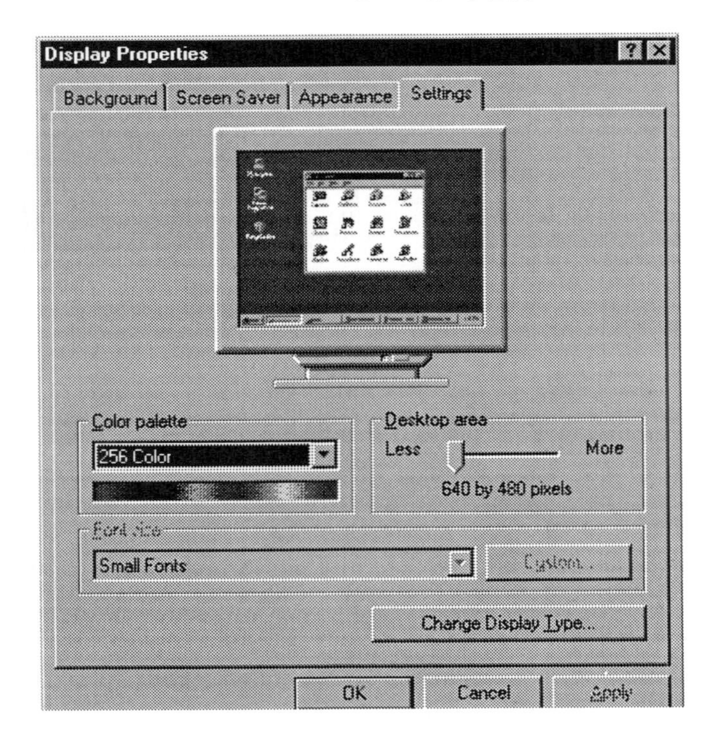

c. In the *Change Display Type* dialog box, click on the **Change** button in the **Monitor Type** frame.

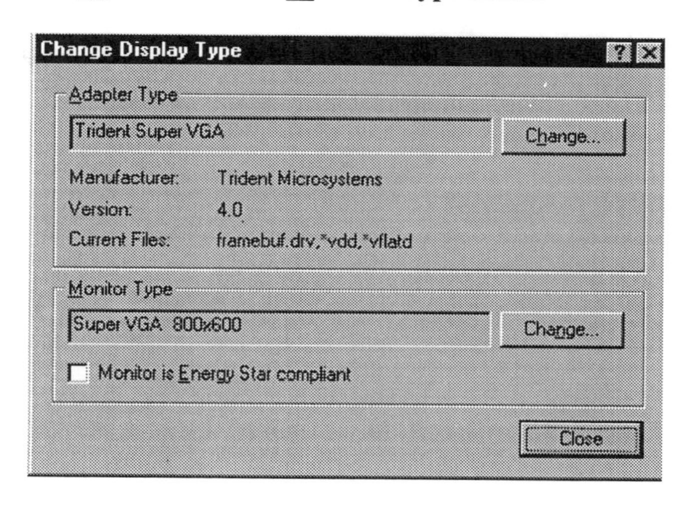

d. In the *Select Device* dialog box, choose the option **Show all devices**, and then select **Standard VGA 640 x 480** in the Models list box.

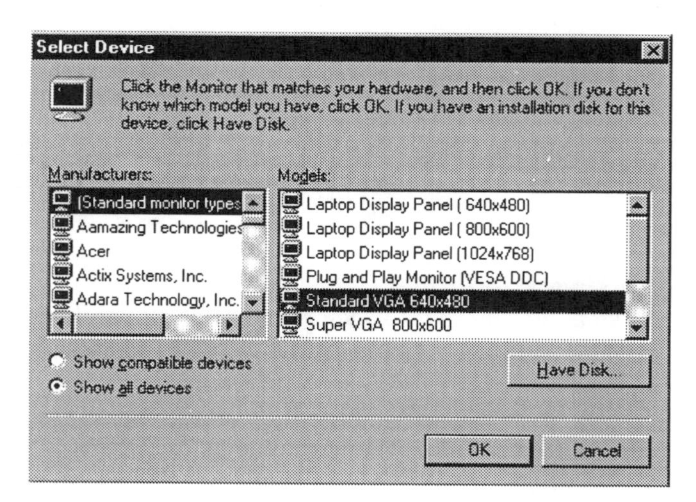

e. Press <Enter> or click on the **OK** button.

Technical Skill Builders

Descriptive Statistics
Students are given a data set and are asked to determine the mean, median, mode, variance, standard deviation and range of the data set.

Discrete Random Variables
Students are given a discrete probability distribution and are asked to determine the expected value, the variance, and the standard deviation of the given distribution, as well as the probability of specific events.

Binomial Word Problems
Word problems are given concerning a Binomial distribution and students are asked to calculate the probability of specific events.

Hypergeometric Word Problems
Word problems are given concerning a hypergeometric distribution and students are asked to calculate the probability of specific events.

Poisson Word Problems
Word problems are given that contain the mean of a Poisson distribution and students are asked to calculate the probability of specific events.

The Standard Normal
Students are asked to determine the probability that a standard normal random variable will lie within some interval.

Normal Distribution Word Problems
Word problems are given containing the mean and the standard deviation of a Normal distribution. Students are asked to calculate the probability that the normal random variable will lie within some interval.

Finding the Value of Z
Students are asked to find the value of z corresponding to a given probability and region for a standard normal distribution.

Sampling Distributions (Means)
Word problems are given containing the mean and standard deviation of a population. The central limit theorem is used to determine the probability that the mean of a random sample will be within some interval.

Sampling Distributions (Proportions)
Word problems are given containing information about a population proportion. Students are asked to solve probability problems concerning the sample proportion.

Hypothesis Testing (Means P Value)
Students are given word problems containing information about a population mean and are asked questions concerning the formulation and testing of a hypothesis. The steps for solving the questions are defining the hypothesis, computing the value of the z test statistic, determining whether the hypothesis should be one or two tailed, entering the p value of the z test statistic, entering the value of the level of significance, and deciding to reject or not to reject the null hypothesis.

Hypothesis Testing (Means Z Value)
Students are given word problems containing information about a population mean and are asked questions concerning the formulation and testing of a hypothesis using the traditional critical value approach.

Hypothesis Testing (Proportions P Value)
Students are given word problems containing information about a population proportion and are asked questions concerning the formulation and testing of a hypothesis. The steps for solving the questions are defining the hypothesis,

computing the value of the z test statistic, determining whether the hypothesis should be one or two tailed, entering the p value of the z test statistic, entering the value of the level of significance, making a decision, and deciding if there is sufficient evidence to support the claim made in the hypothesis.

Hypothesis Testing (Proportions Z Value)
Students are given word problems containing information about a population proportion of a normally distributed population and are asked questions concerning the formulation and testing of a hypothesis. The steps in solving the questions are defining the hypothesis, computing the value of the z test statistic, determining whether the hypothesis should be one or two tailed, determining the critical value of the test statistic, making a decision, and deciding if there is sufficient evidence to support the claim made in the hypothesis.

Fitting a Linear Model
Students are given a set of (x,y) coordinates and are asked to construct the linear model, $y = b_0 + b_1x$ as well as determine various statistics concerning the model.

Regression Analysis I
Given a set of (x,y) coordinates, students are asked to answer questions concerning the estimated model. The five steps to each question are the determination of the sum of squared errors, the variance of errors, the estimated variance of slope, and two confidence intervals for the slope.

Estimation (Means)
Word problems are given containing information concerning the mean and standard deviation of a population. Students are asked to construct confidence intervals and answer sample size determination problems concerning the sample mean.

Estimation (Proportions)
Word problems are given containing information concerning a proportion. Students are asked to construct confidence intervals and answer sample size determination problems concerning the sample proportion.

ANOVA (One Way)
Students are given various parts of an analysis of variance (ANOVA) table and are asked questions regarding the remainder of the table. The first four parts of each question complete the ANOVA table, the fifth part involves the calculation of the sum of squares of sample means about the grand mean, and the sixth part involves the calculation of the variation of the individual measurements about their respective means. In the final two parts of each question, students are asked to find the critical value of the F statistic at a specified level and indicate if that F statistic is significant at the specified level.

ANOVA Regression
This module is similar to ANOVA (One Way), however, the ANOVA table is from a regression analysis. Students are asked to fill in the missing pieces as well as answer various questions regarding inferences that can be made from the table.

Concept Builders

Complete documentation for the following modules can be found in the **Documentation** option under the **Help** option on the module's main menu. The lab exercises for Sampling Distributions and Type II Errors follow these module descriptions.

Name That Distribution
Name That Distribution (NTD) is a concept builder designed to strengthen analytical skills in distribution recognition and data analysis. After considering the available sample information, you must decide the theoretical distribution from which the sample data has been generated. Each time Name That Distribution begins, a different set of game parameters is utilized.

Sampling Distributions
Sampling Distributions is designed to permit experimentation with the sampling distribution of the sample mean. Several experiments have been designed to guide your exploration. We recommend that you read the documentation and then perform Lab Experiment 1 to develop familiarity with the program.

Type II Errors
The Type II Errors Concept Builder is a tool to visually experiment with Type II errors and their probability. Before experimenting with the program or lab exercises you may wish to consult a statistics text to refresh your understanding of Type I and Type II errors.

Sampling Distributions
Lab Experiment 1 **Name** ________________

A. Click on the **<P>lay** option. The simulation initially generates a sample of 30 observations from a uniform distribution [0, 10]. List the 30 observations in the space provided below:

1. ________ 11. ________ 21. ________

2. ________ 12. ________ 22. ________

3. ________ 13. ________ 23. ________

4. ________ 14. ________ 24. ________

5. ________ 15. ________ 25. ________

6. ________ 16. ________ 26. ________

7. ________ 17. ________ 27. ________

8. ________ 18. ________ 28. ________

9. ________ 19. ________ 29. ________

10. ________ 20. ________ 30. ________

Calculate the mean of

(1) the first 5 observations ________

(2) the first 10 observations ________

(3) all 30 observations ________

The values you have calculated should correspond to the means given in the upper left hand corner of the display.

These values are graphically displayed in the corresponding histograms on the right hand side of your screen. The 30 data values are recorded in the histogram labeled "Histogram of Parent" on the left hand side of your screen.

B. Generate another set of 30 observations by selecting the **Next** option. Record the mean of

(1) the first five observations ________

(2) the first ten observations ________

(3) all thirty observations ________

At this point you have generated two samples and have calculated two sample means for each of the different sample sizes.

(4) What is the mean of the two sample means for

n=5________ n=15________ n=30 ________

Verify these values in the data display.

C. Click on the **<R>un** option and generate 10 iterations. One iteration is equivalent to running the **Next** command. When the 10 iterations are completed, each of the sampling histograms will display 10 additional sample means and contain a total of 12 values.

(1) What is the mean of the sample means for the 12 samples of size?

n=5________ n=15 ________ n= 30 ________

(2) What is the relationship between the size of the sample and the mean of the sample means?

(3) What is the variance of the sample means for the 12 samples of size?

n=5_________ n=15________ n=30 _________

(4) What is the relationship between the size of the sample and the variance of the sample means?

(5) What should be the theoretical value for the variance of the sample mean? [Before calculating the variance of the sample means, you must calculate the variance of the distribution you are sampling from. In this case, the distribution is uniform between 0 and 10. The variance of a uniform distribution is $\frac{(b-a)^2}{12}$, where "b" is the maximum value and "a" is the minimum value of the distribution.]

The theoretical variance for samples of size n=30 is _________________.

What is the actual value? ________________

D. What is the apparent distribution of the data displayed in the histogram labeled parent population?

E. Use the **<R>un** command and generate 50 more samples.

(1) Has the relationship between the variance for samples of size n=5, 15, 30 changed?

(2) What is the apparent distribution of the sample means for n=30?

F. The ________________________Theorem states that if a sample of sufficient size is drawn from a population (regardless of its distribution), the sample means will be ______________ distributed with a mean equal to ___________________________. For samples of size n=30, the variance of the sample means will be__________________.

In my experiment, samples of size 5, 15, & 30 were selected from a population that was ______________ distributed. Sample means were calculated from the samples and for the samples of size n=30 the distribution appears to be ________________ distributed with a mean of ______________. For samples of size n=30, the variance was ______________. The theorem states the variance for the samples of size n=30 should be __________.

Sampling Distributions
Lab Experiment 2 **Name** ________________

1. Click on the **Distribution** option from the initial menu or press the <D> key.

Select the Exponential distribution or press <E> <Enter> .

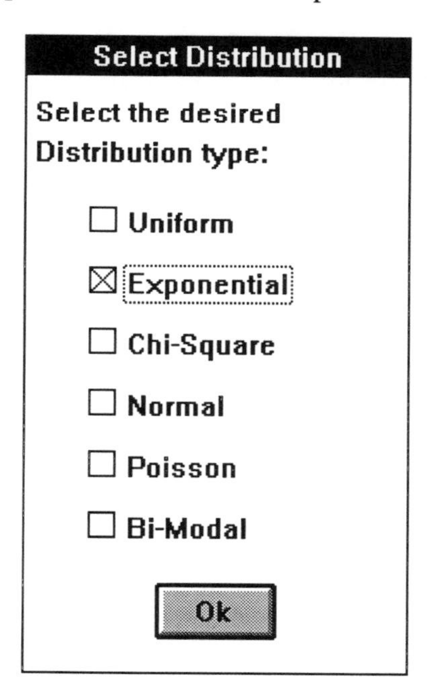

2. Click on the **Play** option or press the <P> key. As in the previous experiment, 30 observations will be generated and three sample means will be computed from the sample data. The sample data is also displayed in the histogram labeled "Parent Histogram".

3. Perform 20 sampling iterations using the **Run** command.

a. What is the apparent distribution of the histogram labeled "Parent Histogram"? ________________

b. What is the mean of the sample means for samples of size

n=5?________ n=15?_________ n=30?________

c. What is the variance of the sample means for

n=5?________ n=15?_________ n=30?________

d. What is the theoretical value of the variance of the sample mean for n=30? __________

4. Use the **Run** command and generate 50 additional sampling iterations.

a. What is the variance of the sample means for samples of size?

n=5?________ n=15?_________ n=30?__________

b. Have the variances changed appreciably since step 3?

c. Sketch the histogram for the sampling distribution of the sample mean for samples of size n=30.

d. Does the distribution appear to be normally distributed?

5. Use the **Run** command to generate 100 more sample iterations.

a. Sketch the distribution for n=30.

b. Does the histogram for n=30 appear normal? If not, generate 100 more sampling iterations.

6. Click on **Done** or press the <D> key. The screen will clear and you will return to the original sampling distribution menu. Change the parent population to bimodal by selecting the **Dist** option on the main menu, then click Bimodal or press <B> & <Enter>.

7. Select the **Play** option. Then select the **Run** option and perform 50 sampling iterations. Because of the increased computational overhead, the iterations will be performed at a slightly slower pace.

a. Sketch the distribution of the parent population.

b. Sketch the distribution of the sample mean for n=30.

c. If the data does not appear normal, generate another 50 sampling iterations and sketch the distribution for n=30.

8. Does the distribution of the parent seem to affect how quickly the distribution of the sample mean becomes normally shaped? Which parent distribution appeared to become normal after the fewest iterations?

____ uniform ____ exponential _____ bimodal

9. Summarize your experimental results in relationship to the Central Limit Theorem.

Type II Errors
Lab Exercise

Name ______________

Click on the **Play** option before beginning the lab exercises.

1. Write down the hypothesis, alpha, sample size, and population variance. This information is provided on your screen.

H_0:	$\alpha =$	$\sigma^2 =$

H_a:	$n =$

2. What is the standard deviation of the sampling distribution of the sample mean?

3. a) Suppose the true population mean is approximately 1/2 standard deviation larger than the hypothesized mean given in the null hypothesis (H_0). What is the value of the population mean? [Hint: be sure to use the standard deviation you found in question 2.]

b) For the population mean you found in part (a), what is the probability of making a type II error?

4. Create a table of population means and determine the probability of making a Type II error for each mean. Use at least 7 different means in your table. Try to select 3 values within 1 standard deviation of the hypothesized mean, 2 values between 1 and 2 standard deviations, and 2 values greater than 2 standard deviations.

True Mean	β

n = _____

5. Plot the values in the table.

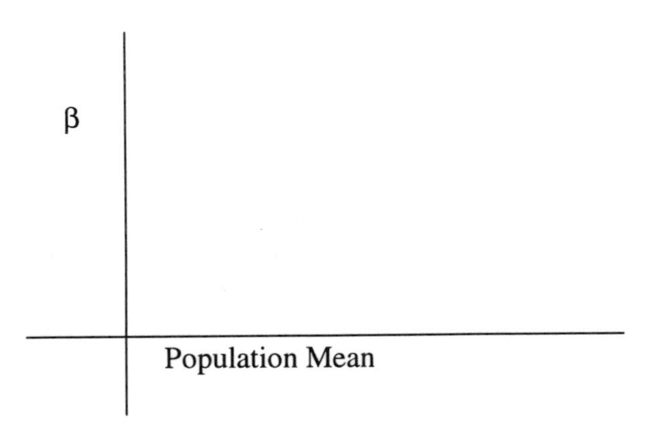

6. Connect the points to form an operating characteristic curve.

7. As the population mean moves further away from the hypothesized mean, the probability of making a Type II error gets ____________ [larger, smaller].

8. If the population mean is very close to the hypothesized mean, Beta will be __________ [larger, smaller] than a typical value of alpha.

9. Suppose the sample size is doubled, use the same population means you used in question [4] to construct new values of Beta.

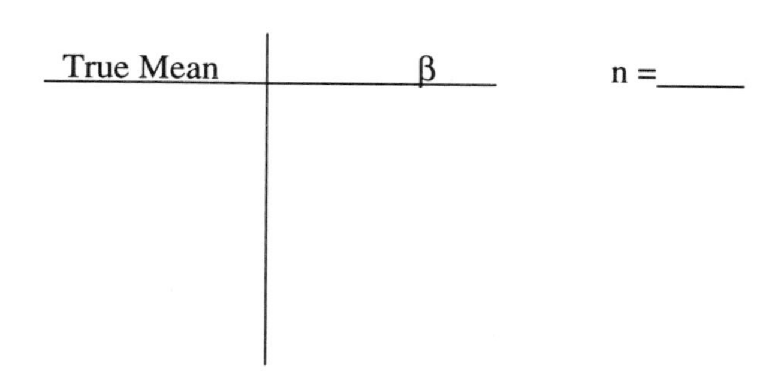

True Mean	β

n =_____

10. Use the points you generated in question [9] to create another OC curve. Draw the OC curve on the graph you used in question [5].

11. What affect does sample size have on the probability of making a Type II error?

12. Given the following hypothesis:

H_o: $\mu \leq 10$

H_a: $\mu > 10$

Is it possible to make a Type II error if the true population mean is 9.0 ? Explain.

Thinking Statistically

Complete documentation for the following modules can be found in the **Documentation** option under the **Help** option on the module's main menu.

Games of Chance

"Just because you deal with risk doesn't mean you must take chances." (Wall Street Journal ad.)

Business decisions involve risk. Purchasing a new piece of machinery, for example, is a wager that sales will continue or increase at a steady rate. If sales decline, then the business must absorb the cost of the equipment and lose the investment (their bet).

The purpose of Games of Chance is to provide an environment for developing a rational approach to analyzing decisions which involve risk. This game involves making decisions based on the expected value of a play of the game.

Direct Mail

The Direct Mail decision simulation enables the decision maker to control the operation of a direct mail marketing company. More importantly, the simulation provides an environment in which statistical concepts can be applied in the course of making business decisions.

Direct Mail
Questions to Consider

Name ______________

1. What is the break even response rate? Does it change for each list?

2. If a 95% confidence interval for the list response rate is completely above the break even percentage, how would you evaluate the risk associated with mailing the list?

3. If a 99% confidence interval for the list response rate is completely above the break even percentage, how would you evaluate the risk associated with the mailing list?

4. Which mailing strategy is the most effective? Design and carry out an experiment to test your hypothesis.

5. What happens to the response rate if a second mailing is made to a list? A third? What is the relationship between response rate and the number of times a list has been mailed?

Binomial Probability Distribution

n	x	.10	.20	.30	.40	.50	n-x
1	0	.9000	.8000	.7000	.6000	.5000	1
	1	.1000	.2000	.3000	.4000	.5000	0
2	0	.8100	.6400	.4900	.3600	.2500	2
	1	.1800	.3200	.4200	.4800	.5000	1
	2	.0100	.0400	.0900	.1600	.2500	0
3	0	.7290	.5120	.3430	.2160	.1250	3
	1	.2430	.3840	.4410	.4320	.3750	2
	2	.0270	.0960	.1890	.2880	.3750	1
	3	.0010	.0080	.0270	.0640	.1250	0
4	0	.6561	.4096	.2401	.1296	.0625	4
	1	.2916	.4096	.4116	.3456	.2500	3
	2	.0486	.1536	.2646	.3456	.3750	2
	3	.0036	.0256	.0756	.1536	.2500	1
	4	.0001	.0016	.0081	.0256	.0625	0
5	0	.5905	.3277	.1681	.0778	.0312	5
	1	.3280	.4096	.3602	.2592	.1562	4
	2	.0729	.2048	.3087	.3456	.3125	3
	3	.0081	.0512	.1323	.2304	.3125	2
	4	.0004	.0064	.0284	.0768	.1562	1
	5	.0000	.0003	.0024	.0102	.0312	0
6	0	.5314	.2621	.1176	.0467	.0156	6
	1	.3543	.3932	.3025	.1866	.0938	5
	2	.0984	.2458	.3241	.3110	.2344	4
	3	.0146	.0819	.1852	.2765	.3125	3
	4	.0012	.0154	.0595	.1382	.2344	2
	5	.0001	.0015	.0102	.0369	.0938	1
	6	.0000	.0001	.0007	.0041	.0156	0
7	0	.4783	.2097	.0824	.0280	.0078	7
	1	.3720	.3670	.2471	.1306	.0547	6
	2	.1240	.2753	.3177	.2613	.1641	5

n	x	.10	.20	.30	.40	.50	n-x
	3	.0230	.1147	.2269	.2903	.2734	4
	4	.0026	.0287	.0972	.1935	.2734	3
	5	.0002	.0043	.0250	.0774	.1641	2
	6	.0000	.0004	.0036	.0172	.0547	1
	7	.0000	.0000	.0002	.0016	.0078	0
8	0	.4305	.1678	.0576	.0168	.0039	8
	1	.3826	.3355	.1977	.0896	.0312	7
	2	.1488	.2936	.2965	.2090	.1094	6
	3	.0331	.1468	.2541	.2787	.2188	5
	4	.0046	.0459	.1361	.2322	.2734	4
	5	.0004	.0092	.0467	.1239	.2188	3
	6	.0000	.0011	.0100	.0413	.1094	2
	7	.0000	.0001	.0012	.0079	.0312	1
	8	.0000	.0000	.0001	.0007	.0039	0
9	0	.3874	.1342	.0404	.0101	.0020	9
	1	.3874	.3020	.1556	.0605	.0176	8
	2	.1722	.3020	.2668	.1612	.0703	7
	3	.0446	.1762	.2668	.2508	.1641	6
	4	.0074	.0661	.1715	.2508	.2461	5
	5	.0008	.0165	.0735	.1672	.2461	4
	6	.0001	.0028	.0210	.0743	.1641	3
	7	.0000	.0003	.0039	.0212	.0703	2
	8	.0000	.0000	.0004	.0035	.0176	1
	9	.0000	.0000	.0000	.0003	.0020	0
10	0	.3487	.1074	.0282	.0060	.0010	10
	1	.3874	.2684	.1211	.0403	.0098	9
	2	.1937	.3020	.2335	.1209	.0439	8
	3	.0574	.2013	.2668	.2150	.1172	7
	4	.0112	.0881	.2001	.2508	.2051	6
	5	.0015	.0264	.1029	.2007	.2461	5
	6	.0001	.0055	.0368	.1115	.2051	4
	7	.0000	.0008	.0090	.0425	.1172	3
	8	.0000	.0001	.0014	.0106	.0439	2
	9	.0000	.0000	.0001	.0016	.0098	1
	10	.0000	.0000	.0000	.0001	.0010	0

n	x	.10	.20	.30	.40	.50	n-x
11	0	.3138	.0859	.0198	.0036	.0005	11
	1	.3835	.2362	.0932	.0266	.0054	10
	2	.2131	.2953	.1998	.0887	.0269	9
	3	.0710	.2215	.2568	.1774	.0806	8
	4	.0158	.1107	.2201	.2365	.1611	7
	5	.0025	.0388	.1321	.2207	.2256	6
	6	.0003	.0097	.0566	.1471	.2256	5
	7	.0000	.0017	.0173	.0701	.1611	4
	8	.0000	.0002	.0037	.0234	.0806	3
	9	.0000	.0000	.0005	.0052	.0269	2
	10	.0000	.0000	.0000	.0007	.0054	1
	11	.0000	.0000	.0000	.0000	.0005	0
12	0	.2824	.0687	.0138	.0022	.0002	12
	1	.3766	.2062	.0712	.0174	.0029	11
	2	.2301	.2835	.1678	.0639	.0161	10
	3	.0852	.2362	.2397	.1419	.0537	9
	4	.0213	.1329	.2311	.2128	.1208	8
	5	.0038	.0532	.1585	.2270	.1934	7
	6	.0005	.0155	.0792	.1766	.2256	6
	7	.0000	.0033	.0291	.1009	.1934	5
	8	.0000	.0005	.0078	.0420	.1208	4
	9	.0000	.0001	.0015	.0125	.0537	3
	10	.0000	.0000	.0002	.0025	.0161	2
	11	.0000	.0000	.0000	.0003	.0029	1
	12	.0000	.0000	.0000	.0000	.0002	0
13	0	.2542	.0550	.0097	.0013	.0001	13
	1	.3672	.1787	.0540	.0113	.0016	12
	2	.2448	.2680	.1388	.0453	.0095	11
	3	.0997	.2457	.2181	.1107	.0349	10
	4	.0277	.1535	.2337	.1845	.0873	9
	5	.0055	.0691	.1803	.2214	.1571	8
	6	.0008	.0230	.1030	.1968	.2095	7
	7	.0001	.0058	.0442	.1312	.2095	6
	8	.0000	.0011	.0142	.0656	.1571	5
	9	.0000	.0001	.0034	.0243	.0873	4
	10	.0000	.0000	.0006	.0065	.0349	3
	11	.0000	.0000	.0001	.0012	.0095	2
	12	.0000	.0000	.0000	.0001	.0016	1
	13	.0000	.0000	.0000	.0000	.0001	0

n	x	.10	.20	.30	.40	.50	n-x
14	0	.2288	.0440	.0068	.0008	.0001	14
	1	.3559	.1539	.0407	.0073	.0009	13
	2	.2570	.2501	.1134	.0317	.0056	12
	3	.1142	.2501	.1943	.0845	.0222	11
	4	.0348	.1720	.2290	.1549	.0611	10
	5	.0078	.0860	.1963	.2066	.1222	9
	6	.0013	.0322	.1262	.2066	.1833	8
	7	.0002	.0092	.0618	.1574	.2095	7
	8	.0000	.0020	.0232	.0918	.1833	6
	9	.0000	.0003	.0066	.0408	.1222	5
	10	.0000	.0000	.0014	.0136	.0611	4
	11	.0000	.0000	.0002	.0033	.0222	3
	12	.0000	.0000	.0000	.0005	.0056	2
	13	.0000	.0000	.0000	.0001	.0009	1
	14	.0000	.0000	.0000	.0000	.0001	0
15	0	.2059	.0352	.0047	.0005	.0000	15
	1	.3432	.1319	.0305	.0047	.0005	14
	2	.2669	.2309	.0916	.0219	.0032	13
	3	.1285	.2501	.1700	.0634	.0139	12
	4	.0428	.1876	.2186	.1268	.0417	11
	5	.0105	.1032	.2061	.1859	.0916	10
	6	.0019	.0430	.1472	.2066	.1527	9
	7	.0003	.0138	.0811	.1771	.1964	8
	8	.0000	.0035	.0348	.1181	.1964	7
	9	.0000	.0007	.0116	.0612	.1527	6
	10	.0000	.0001	.0030	.0245	.0916	5
	11	.0000	.0000	.0006	.0074	.0417	4
	12	.0000	.0000	.0001	.0016	.0139	3
	13	.0000	.0000	.0000	.0003	.0032	2
	14	.0000	.0000	.0000	.0000	.0005	1
	15	.0000	.0000	.0000	.0000	.0000	0
16	0	.1853	.0281	.0033	.0003	.0000	16
	1	.3294	.1126	.0228	.0030	.0002	15
	2	.2745	.2111	.0732	.0150	.0018	14
	3	.1423	.2463	.1465	.0468	.0085	13
	4	.0514	.2001	.2040	.1014	.0278	12
	5	.0137	.1201	.2099	.1623	.0667	11
	6	.0028	.0550	.1649	.1983	.1222	10
	7	.0004	.0197	.1010	.1889	.1746	9

n	x	.10	.20	.30	.40	.50	n-x
	8	.0001	.0055	.0487	.1417	.1964	8
	9	.0000	.0012	.0185	.0840	.1746	7
	10	.0000	.0002	.0056	.0392	.1222	6
	11	.0000	.0000	.0013	.0142	.0667	5
	12	.0000	.0000	.0002	.0040	.0278	4
	13	.0000	.0000	.0000	.0008	.0085	3
	14	.0000	.0000	.0000	.0001	.0018	2
	15	.0000	.0000	.0000	.0000	.0002	1
	16	.0000	.0000	.0000	.0000	.0000	0
17	0	.1668	.0225	.0023	.0002	.0000	17
	1	.3150	.0957	.0169	.0019	.0001	16
	2	.2800	.1914	.0581	.0102	.0010	15
	3	.1556	.2393	.1245	.0341	.0052	14
	4	.0605	.2093	.1868	.0796	.0182	13
	5	.0175	.1361	.2081	.1379	.0472	12
	6	.0039	.0680	.1784	.1839	.0944	11
	7	.0007	.0267	.1201	.1927	.1484	10
	8	.0001	.0084	.0644	.1606	.1855	9
	9	.0000	.0021	.0276	.1070	.1855	8
	10	.0000	.0004	.0095	.0571	.1484	7
	11	.0000	.0001	.0026	.0242	.0944	6
	12	.0000	.0000	.0006	.0081	.0472	5
	13	.0000	.0000	.0001	.0021	.0182	4
	14	.0000	.0000	.0000	.0004	.0052	3
	15	.0000	.0000	.0000	.0001	.0010	2
	16	.0000	.0000	.0000	.0000	.0001	1
	17	.0000	.0000	.0000	.0000	.0000	0
18	0	.1501	.0180	.0016	.0001	.0000	18
	1	.3002	.0811	.0126	.0012	.0001	17
	2	.2835	.1723	.0458	.0069	.0006	16
	3	.1680	.2297	.1046	.0246	.0031	15
	4	.0700	.2153	.1681	.0614	.0117	14
	5	.0218	.1507	.2017	.1146	.0327	13
	6	.0052	.0816	.1873	.1655	.0708	12
	7	.0010	.0350	.1376	.1892	.1214	11
	8	.0002	.0120	.0811	.1734	.1669	10
	9	.0000	.0033	.0386	.1284	.1855	9
	10	.0000	.0008	.0149	.0771	.1669	8
	11	.0000	.0001	.0046	.0374	.1214	7

n	x	.10	.20	.30	.40	.50	n-x
	12	.0000	.0000	.0012	.0145	.0708	6
	13	.0000	.0000	.0002	.0044	.0327	5
	14	.0000	.0000	.0000	.0011	.0117	4
	15	.0000	.0000	.0000	.0002	.0031	3
	16	.0000	.0000	.0000	.0000	.0006	2
	17	.0000	.0000	.0000	.0000	.0001	1
	18	.0000	.0000	.0000	.0000	.0000	0
19	0	.1351	.0144	.0011	.0001	.0000	19
	1	.2852	.0685	.0093	.0008	.0000	18
	2	.2852	.1540	.0358	.0046	.0003	17
	3	.1796	.2182	.0869	.0175	.0018	16
	4	.0798	.2182	.1491	.0467	.0074	15
	5	.0266	.1636	.1916	.0933	.0222	14
	6	.0069	.0955	.1916	.1451	.0518	13
	7	.0014	.0443	.1525	.1797	.0961	12
	8	.0002	.0166	.0981	.1797	.1442	11
	9	.0000	.0051	.0514	.1464	.1762	10
	10	.0000	.0013	.0220	.0976	.1762	9
	11	.0000	.0003	.0077	.0532	.1442	8
	12	.0000	.0000	.0022	.0237	.0961	7
	13	.0000	.0000	.0005	.0085	.0518	6
	14	.0000	.0000	.0001	.0024	.0222	5
	15	.0000	.0000	.0000	.0005	.0074	4
	16	.0000	.0000	.0000	.0001	.0018	3
	17	.0000	.0000	.0000	.0000	.0003	2
	18	.0000	.0000	.0000	.0000	.0000	1
	19	.0000	.0000	.0000	.0000	.0000.	0
20	0	.1216	.0115	.0008	.0000	.0000	20
	1	.2702	.0576	.0068	.0005	.0000	19
	2	.2852	.1369	.0278	.0031	.0002	18
	3	.1901	.2054	.0716	.0123	.0011	17
	4	.0898	.2182	.1304	.0350	.0046	16
	5	.0319	.1746	.1789	.0746	.0148	15
	6	.0089	.1091	.1916	.1244	.0370	14
	7	.0020	.0545	.1643	.1659	.0739	13
	8	.0004	.0222	.1144	.1797	.1201	12
	9	.0001	.0074	.0654	.1597	.1602	11
	10	.0000	.0020	.0308	.1171	.1762	10
	11	.0000	.0005	.0120	.0710	.1602	9

n	x	.10	.20	.30	.40	.50	n-x
	12	.0000	.0001	.0039	.0355	.1201	8
	13	.0000	.0000	.0010	.0146	.0739	7
	14	.0000	.0000	.0002	.0049	.0370	6
	15	.0000	.0000	.0000	.0013	.0148	5
	16	.0000	.0000	.0000	.0003	.0046	4
	17	.0000	.0000	.0000	.0000	.0011	3
	18	.0000	.0000	.0000	.0000	.0002	2
	19	.0000	.0000	.0000	.0000	.0000	1
	20	.0000	.0000	.0000	.0000	.0000	0

Poisson Table

x	1.5	1.6	1.7	1.8	1.9	2.0	2.1	2.2
0	2231	2019	1827	1653	1496	1353	1225	1108
1	3347	3230	3106	2975	2842	2707	2572	2438
2	2510	2584	2640	2678	2700	2707	2700	2681
3	1255	1378	1496	1607	1710	1804	1890	1966
4	0471	0551	0636	0723	0812	0902	0992	1082
5	0141	0176	0216	0260	0309	0361	0417	0476
6	0035	0047	0061	0078	0098	0120	0146	0174
7	0008	0011	0015	0020	0027	0034	0044	0055
8	0001	0002	0003	0005	0006	0009	0011	0015
9	0000	0000	0001	0001	0001	0002	0003	0004
10	0000	0000	0000	0000	0000	0000	0000	0001

x	2.3	2.4	2.5	2.6	2.7	2.8	2.9	3.0
0	1003	0907	0821	0743	0672	0608	0050	0498
1	2306	2177	2052	1931	1815	1703	1596	1494
2	2652	2613	2565	2510	2450	2384	2314	2240
3	2033	2090	2138	2176	2205	2225	2237	2240
4	1169	1254	1336	1414	1488	1557	1622	1680
5	0538	0602	0668	0735	0804	0872	0940	1008
6	0206	0241	0278	0319	0362	0407	0455	0504
7	0068	0083	0099	0118	0139	0163	0188	0216
8	0019	0025	0031	0038	0047	0057	0068	0081
9	0005	0007	0009	0011	0014	0018	0022	0027
10	0001	0002	0002	0003	0004	0005	0006	0008
11	0000	0000	0000	0001	0001	0001	0002	0002
12	0000	0000	0000	0000	0000	0000	0000	0001

x	3.1	3.2	3.3	3.4	3.5	3.6	3.7	3.8
0	0450	0408	0369	0334	0302	0273	0247	0224
1	1397	1304	1217	1135	1057	0984	0915	0850
2	2165	2087	2008	1929	1850	1771	1692	1615
3	2237	2226	2209	2186	2158	2125	2087	2046
4	1734	1781	1823	1858	1888	1912	1931	1944
5	1075	1140	1203	1264	1322	1377	1429	1477
6	0555	0608	0662	0716	0771	0826	0881	0936

x	3.1	3.2	3.3	3.4	3.5	3.6	3.7	3.8
7	0246	0278	0312	0348	0385	0425	0466	0508
8	0095	0111	0129	0148	0169	0191	0215	0241
9	0033	0040	0047	0056	0066	0076	0089	0102
10	0010	0013	0016	0019	0023	0028	0033	0039
11	0003	0004	0005	0006	0007	0009	0011	0013
12	0001	0001	0001	0002	0002	0003	0003	0004
13	0000	0000	0000	0000	0001	0001	0001	0001
14	0000	0000	0000	0000	0000	0000	0000	0000

x	3.9	4.0	4.1	4.2	4.3	4.4	4.5	4.6
0	0202	0183	0166	0150	0136	0123	0111	0101
1	0789	0733	0679	0630	0583	0540	0500	0462
2	1539	1465	1393	1323	1254	1188	1125	1063
3	2001	1954	1904	1852	1798	1743	1687	1631
4	1951	1954	1951	1944	1933	1917	1898	1875
5	1522	1563	1600	1633	1662	1687	1708	1725
6	0989	1042	1093	1143	1191	1237	1281	1323
7	0551	0595	0640	0686	0732	0778	0824	0869
8	0269	0298	0328	0360	0393	0428	0463	0500
9	0116	0132	0150	0168	0188	0209	0232	0255
10	0045	0053	0061	0071	0081	0092	0104	0118
11	0016	0019	0023	0027	0032	0037	0043	0049
12	0005	0006	0008	0009	0011	0014	0016	0019
13	0002	0002	0002	0003	0004	0005	0006	0007
14	0000	0001	0001	0001	0001	0001	0002	0002
15	0000	0000	0000	0000	0000	0000	0001	0001

x	4.7	4.8	4.9	5.0	5.1	5.2	5.3	5.4
0	0091	0082	0074	0067	0061	0055	0050	0045
1	0427	0395	0365	0337	0311	0287	0265	0244
2	1005	0948	0894	0842	0793	0746	0701	0659
3	1574	1517	1460	1404	1348	1293	1239	1185
4	1849	1820	1789	1755	1719	1681	1641	1600
5	1738	1747	1753	1755	1753	1748	1740	1728
6	1362	1398	1432	1462	1490	1515	1537	1555
7	0914	0959	1002	1044	1086	1125	1163	1200

x	4.7	4.8	4.9	5.0	5.1	5.2	5.3	5.4
8	0537	0575	0614	0653	0692	0731	0771	0810
9	0280	0307	0334	0363	0392	0423	0454	0486
10	0132	0147	0164	0181	0200	0220	0241	0262
11	0056	0064	0073	0082	0093	0104	0116	0129
12	0022	0026	0030	0034	0039	0045	0051	0058
13	0008	0009	0011	0013	0015	0018	0021	0024
14	0003	0003	0004	0005	0006	0007	0008	0009
15	0001	0001	0001	0002	0002	0002	0003	0003
16	0000	0000	0000	0000	0001	0001	0001	0001
17	0000	0000	0000	0000	0000	0000	0000	0000

x	5.5	5.6	5.7	5.8	5.9	6.0	6.1	6.2
0	0041	0037	0033	0030	0027	0025	0022	0020
1	0225	0207	0191	0176	0162	0149	0137	0126
2	0618	0580	0544	0509	0477	0446	0417	0390
3	1133	1082	1033	0985	0938	0892	0848	0806
4	1558	1515	1472	1428	1383	1339	1294	1249
5	1714	1697	1678	1656	1632	1606	1579	1549
6	1571	1584	1594	1601	1605	1606	1605	1601
7	1234	1267	1298	1326	1353	1377	1399	1418
8	0849	0887	0925	0962	0998	1033	1066	1099
9	0519	0552	0586	0620	0654	0688	0723	0757
10	0285	0309	0334	0359	0386	0413	0441	0469
11	0143	0157	0173	0190	0207	0225	0245	0265
12	0065	0073	0082	0092	0102	0113	0124	0137
13	0028	0032	0036	0041	0046	0052	0058	0065
14	0011	0013	0015	0017	0019	0022	0025	0029
15	0004	0005	0006	0007	0008	0009	0010	0012
16	0001	0002	0002	0002	0003	0003	0004	0005
17	0000	0001	0001	0001	0001	0001	0001	0002
18	0000	0000	0000	0000	0000	0000	0000	0001
19	0000	0000	0000	0000	0000	0000	0000	0000

x	6.3	6.4	6.5	6.6	6.7	6.8	6.9	7.0
0	0018	0017	0015	0014	0012	0011	0010	0009
1	0116	0106	0098	0090	0082	0076	0070	0064

x	6.3	6.4	6.5	6.6	6.7	6.8	6.9	7.0
2	0364	0340	0318	0296	0276	0258	0240	0223
3	0765	0726	0688	0652	0617	0584	0552	0521
4	1205	1162	1118	1076	1034	0992	0952	0912
5	1519	1487	1454	1420	1385	1349	1314	1277
6	1595	1586	1575	1562	1546	1529	1511	1490
7	1435	1450	1462	1472	1480	1486	1489	1490
8	1130	1160	1188	1215	1240	1263	1284	1304
9	0791	0825	0858	0891	0923	0954	0985	1014
10	0498	0528	0558	0588	0618	0649	0679	0710
11	0285	0307	0330	0353	0377	0401	0426	0452
12	0150	0164	0179	0194	0210	0227	0245	0264
13	0073	0081	0089	0098	0108	0119	0130	0142
14	0033	0037	0041	0046	0052	0058	0064	0071
15	0014	0016	0018	0020	0023	0026	0029	0033
16	0005	0006	0007	0008	0010	0011	0013	0014
17	0002	0002	0003	0003	0004	0004	0005	0006
18	0001	0001	0001	0001	0001	0002	0002	0002
19	0000	0000	0000	0000	0000	0001	0001	0001

x	7.1	7.2	7.3	7.4	7.5	7.6	7.7	7.8
0	0008	0007	0007	0006	0006	0005	0005	0004
1	0059	0054	0049	0045	0041	0038	0035	0032
2	0208	0194	0180	0167	0156	0145	0134	0125
3	0492	0464	0438	0413	0389	0366	0345	0324
4	0874	0836	0799	0764	0729	0696	0663	0632
5	1241	1204	1167	1130	1094	1057	1021	0986
6	1468	1445	1420	1394	1367	1339	1311	1282
7	1489	1486	1481	1474	1465	1454	1442	1428
8	1321	1337	1351	1363	1373	1382	1388	1392
9	1042	1070	1096	1121	1144	1167	1187	1207
10	0740	0770	0800	0829	0858	0887	0914	0941
11	0478	0504	0531	0558	0585	0613	0640	0667
12	0283	0303	0323	0344	0366	0388	0411	0434
13	0154	0168	0181	0196	0211	0227	0243	0260
14	0078	0086	0095	0104	0113	0123	0134	0145

x	7.1	7.2	7.3	7.4	7.5	7.6	7.7	7.8
15	0037	0041	0046	0051	0057	0062	0069	0075
16	0016	0019	0021	0024	0026	0030	0033	0037
17	0007	0008	0009	0010	0012	0013	0015	0017
18	0003	0003	0004	0004	0005	0006	0006	0007
19	0001	0001	0001	0002	0002	0002	0003	0003
20	0000	0000	0001	0001	0001	0001	0001	0001
21	0000	0000	0000	0000	0000	0000	0000	0000

x	7.9	8.0	8.1	8.2	8.3	8.4	8.5	8.6
0	0004	0003	0003	0003	0002	0002	0002	0002
1	0029	0027	0025	0023	0021	0019	0017	0016
2	0116	0107	0100	0092	0086	0079	0074	0068
3	0305	0286	0269	0252	0237	0222	0208	0195
4	0602	0573	0544	0517	0491	0466	0443	0420
5	0951	0916	0882	0849	0816	0784	0752	0772
6	1252	1221	1191	1160	1128	1097	1066	1034
7	1413	1396	1378	1358	1338	1317	1294	1271
8	1395	1396	1395	1392	1388	1382	1375	1366
9	1224	1241	1256	1269	1280	1290	1299	1306
10	0967	0993	1017	1040	1063	1084	1104	1123
11	0695	0722	0749	0776	0802	0828	0853	0878
12	0457	0481	0505	0530	0555	0579	0604	0629
13	0278	0296	0315	0334	0354	0374	0395	0416
14	0157	0169	0182	0196	0210	0225	0240	0256
15	0083	0090	0098	0107	0116	0126	0136	0147
16	0041	0045	0050	0055	0060	0066	0072	0079
17	0019	0021	0024	0026	0029	0033	0036	0040
18	0008	0009	0011	0012	0014	0015	0017	0019
19	0003	0004	0005	0005	0006	0007	0008	0009
20	0001	0002	0002	0002	0002	0003	0003	0004
21	0001	0001	0001	0001	0001	0001	0001	0002
22	0000	0000	0000	0000	0000	0000	0001	0001

x	8.7	8.8	8.9	9.0	9.1	9.2	9.3	9.4
0	0002	0002	0001	0001	0001	0001	0001	0001
1	0014	0013	0012	0011	0010	0009	0009	0008

x	**8.7**	**8.8**	**8.9**	**9.0**	**9.1**	**9.2**	**9.3**	**9.4**
2	0063	0058	0054	0050	0046	0043	0040	0037
3	0183	0171	0160	0150	0140	0131	0123	0115
4	0398	0377	0357	0337	0319	0302	0285	0269
5	0692	0663	0635	0607	0581	0555	0530	0506
6	1003	0972	0941	0911	0881	0851	0822	0793
7	1427	1222	1197	1171	1145	1118	1091	1064
8	1356	1344	1322	1318	1302	1286	1269	1251
9	1311	1315	1317	1318	1317	1315	1311	1306
10	1140	1157	1172	1186	1198	1210	1219	1228
11	0902	0925	0948	0970	0991	1012	1031	1049
12	0654	0679	0703	0728	0752	0776	0799	0822
13	0438	0459	0481	0504	0526	0549	0572	0594
14	0272	0289	0306	0324	0342	0361	0380	0399
15	0158	0169	0182	0194	0208	0221	0235	0250
16	0086	0093	0101	0109	0118	0127	0137	0147
17	0044	0048	0053	0058	0063	0069	0075	0081
18	0021	0024	0026	0029	0032	0035	0039	0042
19	0010	0011	0012	0014	0015	0017	0019	0021
20	0004	0005	0005	0006	0007	0008	0009	0010
21	0002	0002	0002	0003	0003	0003	0004	0004
22	0001	0001	0001	0001	0001	0001	0002	0002
23	0000	0000	0000	0000	0000	0001	0001	0001
24	0000	0000	0000	0000	0000	0000	0000	0000

x	**9.5**	**9.6**	**9.7**	**9.8**	**9.9**	**10.0**	**11.0**	**12.0**
0	0001	0001	0001	0001	0001	0000	0000	0000
1	0007	0007	0006	0005	0005	0005	0002	0001
x	9.5	9.6	9.7	9.8	9.9	10.0	11.0	12.0
2	0034	0031	0029	0027	0025	0023	0010	0004
3	0107	0100	0093	0087	0081	0076	0037	0018
4	0254	0240	0226	0213	0201	0189	0102	0053
5	0483	0460	0439	0418	0398	0378	0224	0127
6	0764	0736	0709	0682	0656	0631	0411	0255
7	1037	1010	0982	0955	0928	0901	0646	0437
8	1232	1212	1191	1170	1148	1126	0888	0655
9	1300	1293	1284	1274	1263	1251	1085	0874

x	9.5	9.6	9.7	9.8	9.9	10.0	11.0	12.0
10	1235	1241	1245	1249	1250	1251	1194	1048
11	1067	1083	1098	1112	1125	1137	1194	1144
12	0844	0866	0888	0908	0928	0948	1094	1144
13	0617	0640	0662	0685	0707	0729	0926	1056
14	0419	0439	0459	0479	0500	0521	0728	0905
15	0265	0281	0297	0313	0330	0347	0534	0724
16	0157	0168	0180	0192	0204	0217	0367	0543
17	0088	0095	0103	0111	0119	0128	0237	0383
18	0046	0051	0055	0060	0065	0071	0145	0256
19	0023	0026	0028	0031	0034	0037	0084	0161
20	0011	0012	0014	0015	0017	0019	0046	0097
21	0005	0006	0006	0007	0008	0009	0024	0055
22	0002	0002	0003	0003	0004	0004	0012	0030
23	0001	0001	0001	0001	0002	0002	0006	0016
24	0000	0000	0000	0001	0001	0001	0003	0008

x	13.0	14.0	15.0	16.0	17.0	18.0	19.0	20.0
0	0000	0000	0000	0000	0000	0000	0000	0000
1	0000	0000	0000	0000	0000	0000	0000	0000
2	0002	0001	0000	0000	0000	0000	0000	0000
3	0008	0004	0002	0001	0000	0000	0000	0000
4	0027	0013	0006	0003	0001	0001	0000	0000
5	0070	0037	0019	0010	0005	0002	0001	0001
6	0152	0087	0048	0026	0014	0007	0004	0002
7	0281	0174	0104	0060	0034	0018	0010	0005
8	0457	0304	0194	0120	0072	0042	0024	0013
9	0661	0473	0324	0213	0135	0083	0050	0029
10	0859	0663	0486	0341	0230	0150	0095	0058
11	1015	0844	0663	0496	0355	0245	0164	0106
12	1099	0984	0829	0661	0504	0368	0259	0176
13	1099	1060	0956	0814	0658	0509	0378	0271
14	1021	1060	1024	0930	0800	0655	0514	0387

Record Book

Module	Due Date	Certification Number
Descriptive Statistics		
Discrete Random Variables		
Binomial Word Problems		
Hypergeometric Word Problems		
Poisson Word Problems		
The Standard Normal		
Normal Distribution Word Problems		
Finding the Value of Z		
Sampling Distributions (Means)		
Sampling Distributions (Proportions)		
Hypothesis Testing (Means P Value)		
Hypothesis Testing (Means Z Value)		
Hypothesis Testing (Proportions P Value)		
Hypothesis Testing (Proportions Z Value)		
Fitting a Linear Model		
Regression Analysis I		
Estimation (Means)		
Estimation (Proportions)		
ANOVA (One Way)		
ANOVA Regression		
Name That Distribution		
Games of Chance		
Direct Mail